MACMILLAN/McGRAW-HILL

Math

Daily Enrich Workbook

Grade 3

The *McGraw-Hill* Companies

Macmillan
McGraw-Hill

Published by Macmillan/McGraw-Hill, of McGraw-Hill Education, a division of The McGraw-Hill Companies, Inc., Two Penn Plaza, New York, New York 10121.

Printed in the United States of America

3 4 5 6 7 8 9 047 09 08 07 06 05

Contents

Name _____

Counting and Number Patterns

Ms. Cook's students made mistakes on their homework.
Cross out the mistakes.
Write the correct number above each mistake.

1. Counting by Twos

_____

14, 16, 19, 20, 22, 25, 26

2. Counting by Fives

_____

10, 15, 20, 30, 32, 35, 45

3. Counting by Tens

_____

10, 25, 30, 40, 45, 55, 70

4. Counting by Threes

_____

12, 15, 17, 21, 25, 27, 31

5. Even Numbers

_____

20, 24, 25, 28, 30, 33, 35

6. Odd Numbers

_____

50, 51, 53, 54, 56, 59, 60

Explore Place Value

Solve each riddle. The answer will be a 3-digit number. Find each answer in a triangle. Write the answer on the line.

1. The sum of the digits is 9. The ones digit is 1. The other two digits have a difference of 2.

What is the number? _____

2. The sum of the digits is 14. The tens digit is 7. The hundreds digit is 3 less than the tens digit. The ones digit is 1 less than the hundreds digit.

What is the number? _____

3. The sum of the digits is 20. The ones digit is 5 less than the hundreds digit. The tens digit is 1 more than the hundreds digit.

What is the number? _____

4. The hundreds digit is the same number as the ones digit. The tens digit is 3 more than the ones digit. The sum of the digits is 12.

What is the number? _____

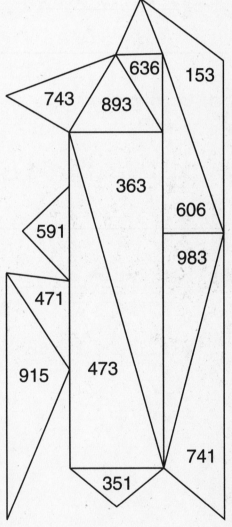

Place Value Through Thousands

Use the block code below to find each number. Write each
number in expanded form and in standard form.

Code: ⬡ = 1,000   ⏢ = 100   ▱ = 10   △ = 1

Number	Expanded Form	Standard Form
1. ⏢⏢⏢⏢ △ △	400 + 2	402
2. ⬡ ▱▱ △△△△△△△		
3. 		
4. 		
5. 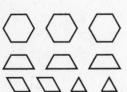		

Place Value Through Hundred Thousands

Use the digits 0 through 9 to answer each question. You may use each digit only one time per question. Write each number in standard form and in word form.

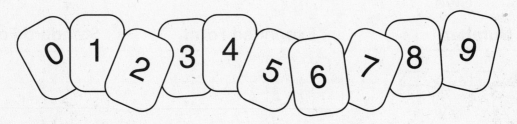

1. What is the greatest 5-digit number with 3 in the tens place? _____

_____

2. What is the greatest 6-digit number with 2 in the ten thousands place? _____

_____

3. What is the least 4-digit number with 0 in the hundreds place? _____

_____

4. What is the least 6-digit number with 9 in the thousands place and 0 in the tens place? _____

_____

5. How does knowing the place value of each digit help you write the word form of a number?

_____

_____

_____

Use with Grade 3, Chapter 1, Lesson 4, pages 10–11.

Name _____

Explore Money

Find the correct change using the fewest bills and coins. Write the number of each coin or bill in the chart. Write the amount of change.

Cost of Game	You Give	🪙	🪙	🪙	🪙	💵	Total Change
Cards $3.95	$5.00						
Marbles $1.39	$2.00						
Jacks $2.25	$5.00						
Checkers $5.76	$10.00						
Chess $7.38	$10.00						
Computer Game $16.57	$20.00						

What strategy did you use to find change using the fewest bills and coins?

_____

_____

Name _____

Count Money and Make Change

Read what each person says. Then find the game or toy that each person bought.

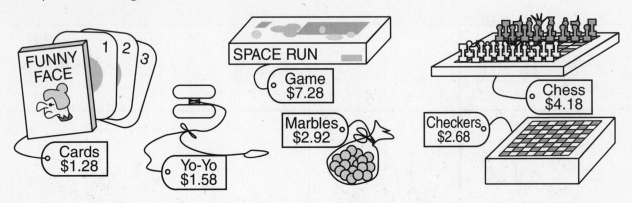

1. "I gave the cashier $5.00. I got back $2.32 in change. What did I buy?"

_____

2. "I got 1 nickel and 3 pennies back in change. I gave the cashier $3.00. What did I buy?"

_____

3. "I gave the cashier 7 quarters and got back $0.17 in change. What did I buy?"

_____

4. "I gave the cashier two $5 bills and got back $2.72 in change. What did I buy?"

_____

5. "I gave the cashier $5.00. I got back two each of four different coins in change. None of the coins are greater than $0.25. What did I buy?"

_____

6. "I gave the cashier two $1 bills. I got back 2 pennies, 2 dimes, and 2 quarters. What did I buy?"

_____

5. How can you check to make sure you were given the right amount of change?

_____

_____

Use with Grade 3, Chapter 1, Lesson 6, pages 14–16.

Problem Solving: Skill
Using the Four-Step Process

Read the problem. Then read each step in the problem-solving process. Write a number next to each step to show the order in which the steps are done.

1. In the first round of a game, Lori won 60 points. Lori won 10 points in each of four more rounds. How many points did Lori win in 5 rounds?

   _____ Identify what you need to find.
   How many points did Lori win in 5 rounds?

   _____ Read the problem. Identify what you know.
   Lori won 60 points in the first round.
   Lori won 10 points in each of four more rounds.

   _____ Make a plan. Start with the points Lori won in the first round. Then count on 10 points for each of the next four rounds.

   _____ Follow your plan to solve the problem. Check your answer.

   How many points did Lori win in 5 rounds? _____

2. Mona has 350 points. Roland has 100 points less than Harvey. Harvey has 200 points more than Mona. Who is the winner?

   _____ Make a plan. Start with Mona's total points. Use her total to find the other totals.

   _____ Identify what you need to find. Who has the most points?

   _____ Use Mona's total to find Harvey's total.
   Harvey has 200 points more than Mona.

   _____ Use Harvey's total to find Roland's total.

   _____ Solve by comparing the numbers. Check your answer.

   _____ Read the problem. Identify what you know.
   Mona has 350 points. Roland has 100 points less than Harvey.
   Harvey has 200 points more than Mona.

   Who is the winner? _____

Compare Numbers and Money

Use the chart to answer the questions below.

World's Largest Animals			
Animal	**Length**	**Height**	**Weight**
Blue Whale	110 feet		
African Bush Elephant		13 feet	
Giraffe		19 feet	
Saltwater Crocodile	16 feet		
Anaconda	27 feet		500 pounds
Whale Shark	41 feet		
Ostrich		9 feet	345 pounds

Solve.

1. Which animal is about twice as long as a saltwater crocodile? _____

2. Which animal weighs more, an anaconda or an ostrich? _____

3. Which animal is longer, a blue whale or an anaconda? _____

4. What length is the longest animal? _____

5. What length is the shortest animal? _____

6. What height is the tallest animal? _____

Problem Solving
Solve.

7. Anna spent $7.95 on a book about insects.
Jane spent $9.75 on a book about fish. Who spent more money?
How can you tell?

_____

_____

Order Numbers and Money

Make as many 3-digit numbers as you can for each set of cards.
Then write your numbers in order.

1.

_____ , _____ , _____

_____ , _____ , _____

Order from least to greatest. _____

2.

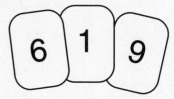

_____ , _____ , _____

_____ , _____ , _____

Order from greatest to least. _____

Make as many money amounts as you can from the set of cards shown below.
Use dollar signs and decimal points. Write amounts in order from least to
greatest.

3.

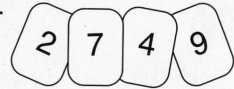

_____

_____

_____

_____

Estimate Quantities

Use the benchmark numbers to estimate the number of objects in each jar.

1. 10 bingo chips

_____

2. 50 pieces of macaroni

_____

3. 25 pennies

_____

4. 5 dog tags

_____

5. 20 marbles

_____

6. 30 dog biscuits

_____

Use with Grade 3, Chapter 2, Lesson 3, pages 30–31.

Name _____

Round to Tens and Hundreds

Read the clues to find the number.

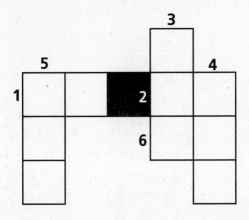

ACROSS

1. When I am rounded to the nearest ten, I am 70. The digit in my ones place is 3 more than the digit in my tens place. What number am I?

2. When I am rounded to the nearest ten, I am 80. The sum of my two digits is 15. What number am I?

6. When I am rounded to the nearest ten, I am 30. The sum of my two digits is 5. What number am I?

DOWN

3. When I am rounded to the nearest hundred, I am 400. The digit in my tens place is 4 more than the digit in my hundreds place. The digits in my hundreds place and ones place are the same. What number am I?

4. When I am rounded to the nearest hundred, I am 800. The digit in my tens place is 6 less than the digit in my hundreds place. The digits in my tens place and ones place are the same. What number am I?

5. When I am rounded to the nearest hundred, I am 700. The digit in my hundreds place is 1 less than the digit in my tens place. My ones digit is 0. The sum of my three digits is 13. What number am I?

Round to the Nearest Thousand

Beth and her friends played a game.
Their scores were rounded to the nearest thousand.

HANA — about 5,000 points

JAMES — about 6,000 points

ALPHONSE — about 4,000 points

JOANN — about 2,000 points

RALPH — about 3,000 points

DEBBIE — about 7,000 points

Each player's actual number of points is listed below.
Match each player's estimated score with the actual score.
Write the player's name on the line.

1. 6,292 points   **2.** 4,368 points   **3.** 6,750 points

_____   _____   _____

4. 4,801 points   **5.** 2,423 points   **6.** 2,500 points

_____   _____   _____

Use with Grade 3, Chapter 2, Lesson 5, pages 34–35.

Addition Properties · Algebra

Play this game with a partner. Take turns.
You will need a number cube labeled 1–6 and two markers.

- Start from the space labeled, START.

- Toss the number cube. Move that number of spaces from left to right.

- Find the sum. Then move **forward** to the closest square that has the same sum. (When you come to the end of a row, move to the beginning of the next row.) Stay in the same place when you cannot find a square with the same sum.

- The first player to reach the END wins.

START →	6 + 5 =	3 + 9 =	8 + 5 =	4 + 7 =	0 + 8 =
3 + 7 =	7 + 6 =	5 + 8 =	9 + 3 =	5 + 9 =	11 + 0 =
5 + 6 =	8 + 0 =	7 + 3 =	8 + 7 =	2 + 3 =	7 + 9 =
13 + 4 =	3 + 6 =	4 + 6 =	9 + 6 =	8 + 8 =	5 + 0 =
0 + 10 =	7 + 5 =	9 + 7 =	8 + 9 =	6 + 3 =	10 + 0 =
7 + 7 =	6 + 4 =	7 + 8 =	6 + 9 =	0 + 5 =	END

Addition Patterns • Algebra

What score does each dartboard show? Add to find the total.
Show the addition sentence you used to find the total.

1.

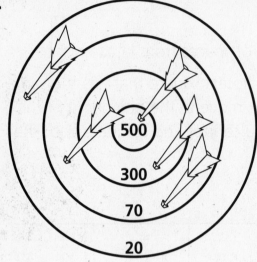

500
300
70
20

_____

2.

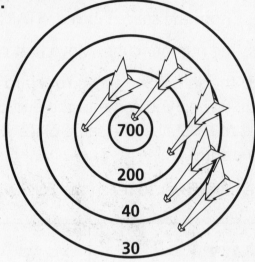

700
200
40
30

_____

3.

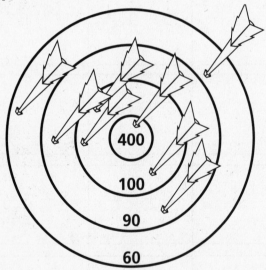

400
100
90
60

_____

4.

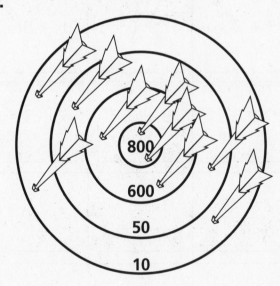

800
600
50
10

_____

5. Look at the addition sentences you wrote. Can you order and
group the numbers in a different way? Tell why or why not.

_____

_____

Explore Regrouping in Addition

Draw the missing models.

1.

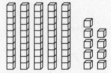

+ _____

The sum is 91.

2.

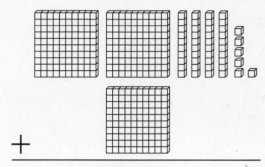

+ _____

The sum is 375.

3.

+ _____

The sum is 112.

4.

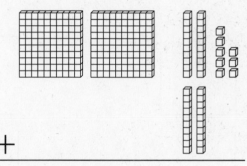

+ _____

The sum is 555.

5.

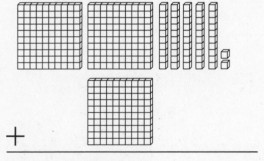

+ _____

The sum is 425.

6.

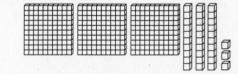

+ _____

The sum is 500.

7. How did you find the missing models in problem 4?

_____

_____

_____

Add Whole Numbers

Use the parts of this figure to decide which number each symbol stands for. Then find the sums. The first one has been done for you.

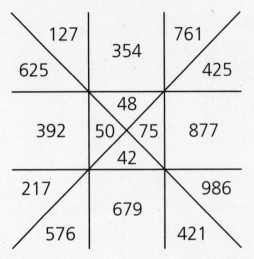

127   761
354
625   425
48
392   50 75   877
42
217   986
679
576   421

1.
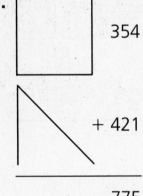

354

+ 421

775

2.
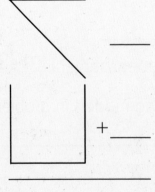

___

+ ___

___

3.

___

+ ___

___

4.

___

+ ___

___

5.

___

+ ___

___

6.

___

+ ___

___

Estimate Sums

Show two different ways to estimate each sum. Show how
you rounded the addends. Then find the exact sum.

1. 128 + 69 = _____

$\underline{\quad100\quad}$ + $\underline{\quad100\quad}$ = _____

$\underline{\quad130\quad}$ + $\underline{\quad70\quad}$ = _____

2. 471 + 35 = _____

_____ + _____ = _____

_____ + _____ = _____

3. 642 + 87 = _____

_____ + _____ = _____

_____ + _____ = _____

4. 893 + 1,329 = _____

_____ + _____ = _____

_____ + _____ = _____

5. 6,321 + 415 = _____

_____ + _____ = _____

_____ + _____ = _____

6. 4,305 + 2,501 = _____

_____ + _____ = _____

_____ + _____ = _____

7. 15,362 + 3,095 = _____

_____ + _____ = _____

_____ + _____ = _____

8. 62,225 + 25,374 = _____

_____ + _____ = _____

_____ + _____ = _____

9. Which way gives you the better estimate? Use one of the problems
as an example. Explain your reasoning.

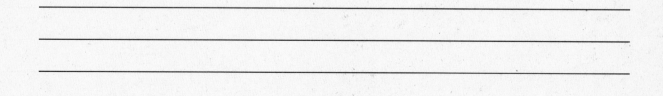

Problem Solving: Skill
Estimate or Exact Answer

Solve. Tell why you decided whether to estimate or find an exact answer.

1. Two school buses arrive at Spruce Farm. One school bus has 49 children. The other school bus has 42 children. Do more than 100 children arrive at Spruce Farm?

_____

_____

2. Lakeside Animal Shelter took in 49 dogs in January, 43 dogs in February, and 37 dogs in March. Did the shelter take in more than 100 dogs during January, February, and March?

_____

_____

_____

3. Spruce Farm has 82 cows and 96 horses. How many horses and cows are at Spruce Farm?

_____

_____

_____

4. Birdland Park has 28 toucans and 47 parrots. How many toucans and parrots are at Birdland Park altogether?

_____

_____

_____

5. Write and solve a problem that needs an estimated answer.

_____

_____

_____

_____

6. Write and solve a problem that needs an exact answer.

_____

_____

_____

_____

Add Greater Numbers

Some numbers have the same value when written forward and backward. They are called *palindromes*. The number **7,997** is a palindrome.

Read forward = 7,997

Read backward = 7997

Start with any number.	1,682
Reverse it.	+ 2,861
Add the numbers.	4,543
Reverse the sum.	+ 3,454
Add the numbers.	7,997

Find the palindrome by starting with each number. Reverse and add. Repeat until you get a palindrome.

Find the palindrome for each number. Show your work.

1. 134 _____   **2.** 328 _____   **3.** 651 _____   **4.** 789 _____

5. 1,234 _____   **6.** 3,571 _____   **7.** 3,099 _____   **8.** 8,957 _____

9. Is this statement true or false? You will always get a palindrome sum when you add two palindrome numbers. Explain.

_____

_____

_____

_____

Add More Than Two Numbers

Look at this magic square.
Find the sum of each row,
column, and diagonal.

_____

Shade the four squares in the
center of the magic square—
356, 355, 360, and 359. What
is the sum of the shaded part
of the magic square?

_____

353	363	364	350
358	356	355	361
354	360	359	357
365	351	352	362

Four other sets of numbers in the magic square have the same sum as the
shaded part. Find the numbers and write them in the squares below.

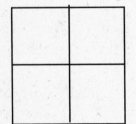   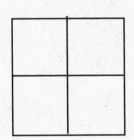   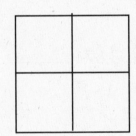   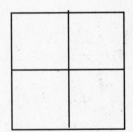

Sum = _____   Sum = _____   Sum = _____   Sum = _____

What strategy did you use to find the four numbers with the same sum?

_____

_____

_____

Use with Grade 3, Chapter 4, Lesson 2, pages 76–77.

Choose a Computation Method

Riddle: How can you make seven an even number?

To solve the riddle, use mental math, paper and pencil, or calculator.

Match each number under the lines with a sum from the problems below. Write the letter next to each sum on the line above the matching number. Watch out! Some sums are not used in the riddle.

___ ___ ___ ___ ___ ___ ___ ___ ___ ___ ___ ___
5,459 9,147 6,645 10,507 9,147 8,249 9,147 3,925 5,459 6,974 10,507 1,699

Add.

1. 3,456 + 2,003 = _____
T

2. 4,962 + 3,287 = _____
W

3. 5,664 + 1,310 = _____
H

4. 2,945 + 1,013 = _____
L

5. 1,800 + 7,347 = _____
A

6. 3,280 + 6,423 = _____
C

7. 6,932 + 216 = _____
G

8. 4,762 + 5,745= _____
E

9. 3,821 + 104 = _____
Y

10. 643 + 332+ 300 = _____
U

11. 1,672 + 200 + 4,773 = _____
K

12. 555 + 769 + 375 = _____
S

13. 6,873 + 4,073 + 863 = _____
B

Name _____

Relate Addition and Subtraction • Algebra

Use the clue to find the addends for each problem.
Then complete the fact family.

1. Clue: The difference between the
addends is 1.

☐ + ☐ = 15

☐ + ☐ = ☐

☐ – ☐ = ☐

☐ – ☐ = ☐

2. Clue: The difference between the
addends is 4.

☐ + ☐ = 14

☐ + ☐ = ☐

☐ – ☐ = ☐

☐ – ☐ = ☐

3. Clue: The difference between the
addends is 3.

☐ + ☐ = 15

☐ + ☐ = ☐

☐ – ☐ = ☐

☐ – ☐ = ☐

4. Clue: The difference between the
addends is 7.

☐ + ☐ = 7

☐ + ☐ = ☐

☐ – ☐ = ☐

☐ – ☐ = ☐

5. Clue: The difference between the
addends is 0.

☐ + ☐ = 18

☐ – ☐ = ☐

6. Clue: The difference between the
addends is 4.

☐ + ☐ = 12

☐ + ☐ = ☐

☐ – ☐ = ☐

☐ – ☐ = ☐

Problem Solving: Skill
Identify Extra Information

Cross out any information that you do not need to solve the problem. Then solve the problem.

1. Howard works for 5 hours at the flower shop. He makes 18 deliveries. This is 6 more deliveries than he made yesterday. How many deliveries did Howard make yesterday?

_____

2. There are 36 children in the garden club. The garden club meets 2 times per month. Each meeting lasts 8 hours. How many hours does the garden club meet each month?

_____

3. Vanna buys 24 roses. The flowers cost $12. Of those roses, 7 are yellow. The rest are red. How many of the roses are red?

_____

4. The garden club plants 9 cherry trees, 16 maple trees, and 15 bushes. How many trees does the garden club plant in all?

_____

Use the poster to answer the questions. Then cross out information in the poster that you do not need to solve the problems.

Saturday, Sept. 5
12:30 PM to 5:30 PM

Garden Club
Picnic

Bring 2 seed packs to share.

Raffle Tickets,
$2 each

Rain Date: Sunday, Sept. 6

5. How long does the picnic last?

_____

6. Seventy-five people come to the picnic. Each one of them brings packs of seeds to share. How many packs of seeds were there all together?

_____

7. Jessie sells 30 raffle tickets before the picnic. How much money did she collect?

_____

Subtraction Patterns • Algebra

Each difference stands for one word. Use the word chart below
to write each riddle and its answer.

Riddle

14 − 8 _____	140 − 80 _____	10 − 9 _____	100 − 90 _____	1,000 − 900 _____?

Answer

15 − 8 _____	150 − 60 _____	150 − 90 _____.

Riddle

15 − 9 _____	130 − 70 _____	1,500 − 900 _____	17 − 9 _____	170 − 90 _____	1,700 − 900 _____?

Answer

11 − 4 _____	110 − 40 _____	110 − 50 _____.

Word Chart

1 wags	7 A	10 its	70 beech	90 dogwood	600 grows
6 What	8 near	60 tree	80 the	100 tail	800 ocean

Use with Grade 3, Chapter 5, Lesson 3, pages 102–103.

Name _____

Explore Regrouping in Subtraction

Play a number game with 1 or 2 players.

You will need place-value models (8 hundreds, 10 tens, 20 ones), construction paper or tagboard, a pencil, and a paper clip.

How to Play

- Spin the spinner three times. Write each number in order to make a 3-digit number.

- Spin the spinner three times again. Write that 3-digit number.

- Use the place-value models to show the greater number.

- Take away the lesser number from the place-value models.

- Write the subtraction problem.

- If there are two players, take turns. Also check each other's subtraction problems. Score 1 point for each correct answer.

- Play at least five rounds of the game.

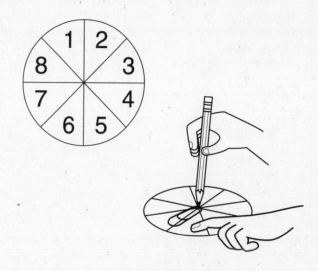

$$\begin{array}{r} {\scriptstyle 2\ 15} \\ 4\cancel{3}\cancel{5} \\ -\ 217 \\ \hline \end{array}$$

How can you check the answer to a subtraction using models?

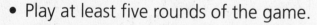

_____

_____

_____

Subtract Whole Numbers

Find each difference to complete the mystery words below.
Match the difference in each box with a number under the lines.
Write the letter in the box on the line.

T 386 − 299	A 152 − 109	T 345 − 258	S 722 − 227	B 543 − 484
R 974 − 185	U 425 − 316	T 587 − 500	I 837 − 508	C 294 − 149

Message

_____ _____ _____ _____ _____ _____ _____ _____   _____ _____!
495   109   59   87   789   43   145   87   329   87

F 372 − 282	I 689 − 122	E 175 − 146	N 423 − 287	E 316 − 287
R 921 − 349	D 231 − 169	E 597 − 568	F 249 − 159	C 731 − 548

Message

_____ _____ _____ _____ _____ _____ _____ _____ _____ _____!
62   567   90   90   29   572   29   136   183   29

Use with Grade 3, Chapter 5, Lesson 5, pages 106–108.

Regroup Across Zeros

Find the missing numbers in each subtraction wheel.

1.

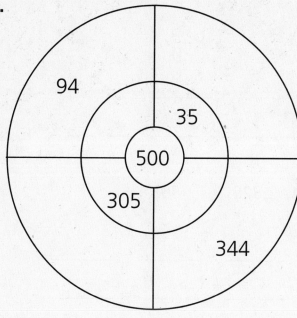

2.

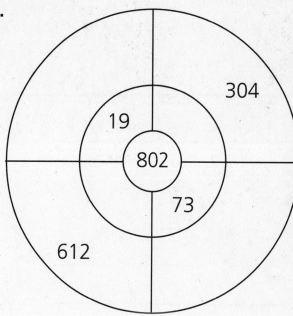

3.

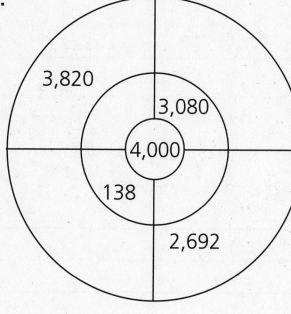

4.

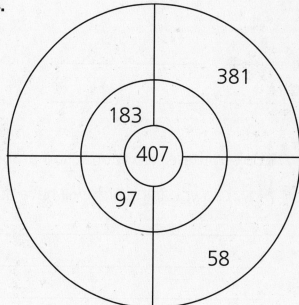

Estimate Differences

Estimate in two different ways. Show how you rounded
the numbers.

1. 167 – 78
  ↓   ↓

$$200 - 100 = 100$$
$$170 - 80 = 90$$

2. 371 – 45
  ↓   ↓

_____ – _____ = _____

_____ – _____ = _____

3. 692 – 82
  ↓   ↓

_____ – _____ = _____

_____ – _____ = _____

4. 2,329 – 893
   ↓   ↓

_____ – _____ = _____

_____ – _____ = _____

5. 7,607 – 943
   ↓   ↓

_____ – _____ = _____

_____ – _____ = _____

6. 4,412 – 315
   ↓   ↓

_____ – _____ = _____

_____ – _____ = _____

7. Look at these two subtraction problems: 739 – 253 and 753 – 239.

How can you tell which will have an exact answer less than 500?

_____

_____

_____

_____

_____

Use with Grade 3, Chapter 6, Lesson 2, pages 116–117.

Subtract Greater Numbers

Complete the subtraction squares in exercises 1–3.
Subtract across and down to find each difference.

You will make your own subtraction square for exercise 4.
Use the numbers at the bottom of the worksheet. Put the
numbers in the right places so that you can subtract across
and down to find the differences.

1.

9,876	5,602	
6,281	2,349	

2.

7,583	4,009	
6,999	3,876	

3.

8,910	6,855	
5,289	4,000	

4.

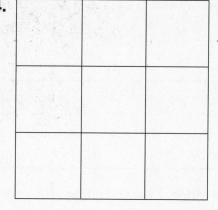

| 9,206 | 3,210 | 5,328 | 2,328 | 2,118 | 3,878 | 3,668 | 210 | 6,878 |

Choose a Computation Method

Choose two numbers from the lists to write 8 subtraction problems in the boxes below. Cross out the numbers you used. Then find each difference.

6,100	9,451	3,982	8,562
8,424	7,601	5,667	7,618
2,197	3,000	1,801	500
5,476	1,283	4,275	6,230

1.	**2.**
3.	**4.**
5.	**6.**
7.	**8.**

If you used mental math, circle the problem in red.
If you used a calculator, circle the problem in green.
If you used paper and pencil, circle the problem in blue.

Which computation method did you use most often? Why?

_____

Use with Grade 3, Chapter 6, Lesson 5, pages 124–125.

Tell Time

Some clocks and watches do not have any numbers.
Look at the clockfaces below. Write the time to the
nearest five minutes.

1.

_____

2.

_____

3.

_____

4.

_____

5.

_____

6.

_____

7.

_____

8.

_____

9.

_____

How can you tell time when a clockface has no numbers?

_____

_____

_____

Name_____

Convert Time

Play a game with a partner.
Take turns.

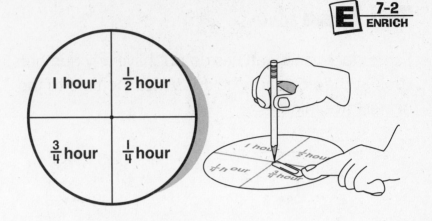

How to Play

- Spin. Write the time in minutes in the chart below.
  For example, if the spinner lands on $\frac{1}{4}$ hour, the player writes
  15 minutes.

- For each turn, add the number of minutes to the previous
  number of minutes. For example, if a player has a total of
  120 minutes and spins $\frac{1}{2}$ hour, then 30 minutes is added to
  120 minutes for a total of 150 minutes.

- After five rounds, check each other's total number of minutes.
  Convert the minutes to hours.

- The player with the greater amount of time wins the game.

Name_____		
Round	Time Spun	Total Time
Round 1		
Round 2		
Round 3		
Round 4		
Round 5		

Name_____		
Round	Time Spun	Total Time
Round 1		
Round 2		
Round 3		
Round 4		
Round 5		

How can knowing how to convert between minutes and hours help you?

_____

_____

Use with Grade 3, Chapter 7, Lesson 2, pages 144–145.

Elapsed Time

Complete this school schedule. There are 5 minutes between classes and activities. Write each time using A.M. or P.M.

Subject	Begin	End	Elapsed Time
Reading	8:35 A.M.	9:25 A.M.	_____
Art	9:30 A.M.	_____	50 minutes
Recess	10:25 A.M.	10:45 A.M.	_____
Math	10:50 A.M.	11:40 A.M.	_____
Lunch	11:45 A.M.	_____	30 minutes
Science	12:20 P.M.	_____	50 minutes
Gym	_____	2:05 P.M.	50 minutes
Social Studies	2:10 P.M.	_____	50 minutes
Dismissal	3:00 P.M.		

1. Starr takes a drum lesson after school. The lesson begins at 3:15 P.M. The lesson lasts for 55 minutes. At what time is the lesson over? _____

2. Dave rides his bicycle home from school. He leaves school at 3:10 P.M. He gets home at 3:47 P.M. How long does Dave ride his bicycle? _____

3. Lisa's alarm clock didn't go off! She was 15 minutes late for her first class. At what time did Lisa get to class? Which class was it? _____

4. Rodrigo twisted his ankle in Gym. He was 20 minutes late for his next class. At what time did he get to class? Which class was it? _____

Calendar

Complete the calendar. Then use it to solve each problem.

November						
S	**M**	**T**	**W**	**T**	**F**	**S**
		1	2	3	4	5
___	7	___	___		11	12
13	___	___	16	___	___	___
___	___	23	24		___	
___	___	___	___	___	___	

1. Election Day is on the first Tuesday after the first Monday of November. Write "Election Day" on the calendar. What is the date?

_____

2. Veteran's Day is 3 days after Election Day this year. Write "Veteran's Day" on the calendar. What is the date?

_____

3. Draw an X on the third Tuesday. What is the date?

_____

4. On which day is the twenty-fourth?

_____

5. Draw a circle on the fourth Thursday. Then label it Thanksgiving Day. What is the date?

_____

6. Erik must water his plants every six days starting on November 3. On which dates must he water the plants?

_____

Time Lines

Ray created a time line to show some important events in his life. Use the data below to fill in the time line.

Read the time line from left to right.

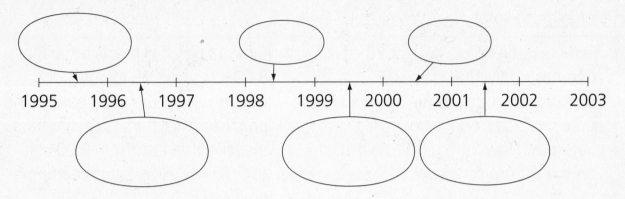

1. Ray got a dog in 1998.

2. He moved from New York City to Maine the year before 1996.

3. Ray's friend came to visit the year after Ray got a dog.

4. Ray's parents bought him a pony in April of 2000.

5. Ray went on a trip to New York City during the summer of 2001.

6. Ray's sister was born the year after he moved to Maine.

Write a time line for some important events in your life.

_____

Problem Solving: Skill
Identify Missing Information

Tell what important information is missing from the problem. If you can find the missing information, solve the problem. If you cannot find the missing information, write "not enough information to solve."

1. Ted works on an art project 20 minutes on Wednesday, 30 minutes on Thursday, until dinner time on Friday, and 1 hour on Saturday. How long did Ted work on the art project in all?

_____

_____

2. Ramon's baseball practice starts at 8:00 A.M., with 15 minutes of running. Then the team has batting practice. After that, the team has a meeting that lasts from 9:00 A.M. to 9:30 A.M. How long is batting practice?

_____

_____

3. There are 6 new members in Rob's game club. How many members does the club have?

_____

_____

4. Max leaves his house at 2:30 P.M. and takes 15 minutes to walk to the mall. He leaves the mall at 4:00 P.M. How much time does he spend at the mall?

_____

_____

5. Ronnie goes out to mow the lawn at 3:45 P.M. She mows for 30 minutes. When she finishes the lawn, she starts her homework. She does homework until 6:00 P.M. How much time does she spend on homework?

_____

_____

6. On Wednesday afternoon, Melanie works 1 hour on science homework and 45 minutes on English homework. The rest of the afternoon, she works on math. How much time did she spend on homework?

_____

_____

Use with Grade 3, Chapter 7, Lesson 6, pages 154–155.

Collect and Organize Data

Survey your classmates to find out what time they go to sleep.
Make a tally chart.
Make a line plot.

Time Students Go to Sleep		
Time	Tally	Number of Students
7:30 P.M.		
8:00 P.M.		
8:30 P.M.		
9:00 P.M.		
9:30 P.M.		
10:00 P.M.		

Time Students Go to Sleep

7:30 8:00 8:30 9:00 9:30 10:00

Use the data to answer problems 1–4.

1. At what time do the most number of students go to sleep?

_____

2. Which two times do the most number of students go to sleep?

_____

3. How many students go to sleep at 9:00 P.M.?

_____

4. Which two times do the fewest number of students go to sleep?

_____

5. How many students go to sleep at 10:00 P.M.?

_____

6. How many students go to sleep the same time as you?

_____

Find Median, Mode, and Range

Complete. Use what you know about the median, mode, and range.

Find the number you need to complete the data sets so the statements are true.

1. The range is 4.   4, 5, 2, 3, 3, 2, 5, 4, _____

2. The median is 5.   6, 7, 3, 2, 3, 6, 6, 2, _____

3. The mode is 2.   5, 1, 4, 2, 5, 2, 3, 1, _____

4. The range is 8.   2, 7, 4, 3, 4, 8, 1, 6, _____

5. The mode is 1.   3, 1, 2, 3, 1, _____

Find each for the given data.

6. mode   2, 3, 4, 6, 5, 3, 3, 2, 5 _____

7. range   5, 9, 8, 5, 9, 4, 5, 8, 6, 4 _____

8. mode   4, 8, 2, 6, 2, 8, 5 ,3, 8, 5 _____

9. median   9, 7, 8, 6, 7, 9, 6, 8, 7 _____

E	I	P	L
1	2	3	5

A	T	O	N
6	7	8	9

Match each answer to problems 1–9 with a number in the cards above. Write the letter in the box above the problem number below. You will spell the answer to the question:

What can help you organize data?

____ ____ ____ ____ ____ ____ ____ ____ ____
  1   2   3   4   5   6   7   8   9

Use with Grade 3, Chapter 8, Lesson 2, pages 162–164.

Explore Pictographs

Read the tongue twister below.

> Peter Piper picked 12 pecks of red pickled peppers. Penny Piper picked 16 pecks of red pickled peppers. Patsy Piper picked 11 pecks of red pickled peppers. Paul Piper picked 13 pecks of red pickled peppers. Perry Piper picked 21 pecks of red pickled peppers.

1. Make a tally chart that shows how many peppers the Pipers picked.

2. Then use the tally chart to help you make a pictograph that shows how many peppers the Pipers picked. Use whole and half symbols in your pictograph.

Choose a symbol. What is it? Draw it here.

_____

How many peppers will each symbol stand for?

_____

Peppers Picked by the Pipers		
Piper	**Tally**	**Number of Peppers**
Peter		
Penny		
Patsy		
Paul		
Perry		

Peppers Picked by the Pipers	
Peter	
Penny	
Patsy	
Paul	
Perry	

Key: Each ✿ stands for _____ peppers.

Who picked the most red pickled peppers? _____

Explore Bar Graphs

It is Travel Day in the third-grade class at Becker Elementary School. Guests have been invited to talk about the places they've visited. The chart and line plot below show how many students attended each talk.

Travel Day Speakers	
Country	**Number of Students Who Attended**
Spain	10
Ghana	11
China	12
New Zealand	14

Students Who Attended Travel Day

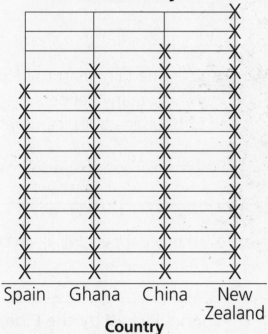

Use the chart and line plot to complete the bar graph.

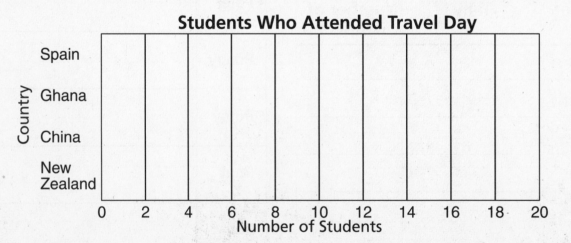

Which way do you think is the best way to show the data? Tell why.

_____

_____

Coordinate Graphs

Use the ordered pairs to find the points on each grid. Then connect the points in the order given. Write the name of each shape that you make.

1.

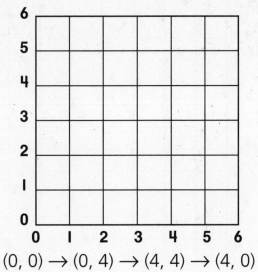

$(0, 0) \rightarrow (0, 4) \rightarrow (4, 4) \rightarrow (4, 0)$

Shape: _____

2.

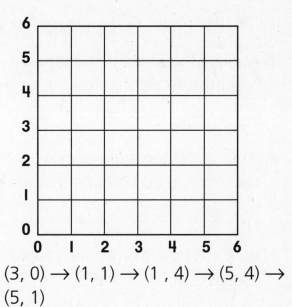

$(3, 0) \rightarrow (1, 1) \rightarrow (1 , 4) \rightarrow (5, 4) \rightarrow (5, 1)$

Shape: _____

3.

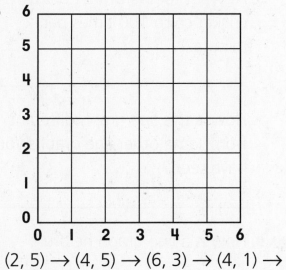

$(2, 5) \rightarrow (4, 5) \rightarrow (6, 3) \rightarrow (4, 1) \rightarrow (2, 1) \rightarrow (0, 3)$

Shape: _____

4.

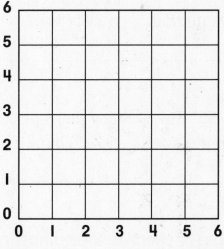

$(3, 6) \rightarrow (4, 4) \rightarrow (6, 4) \rightarrow (4, 3) \rightarrow (5, 1) \rightarrow (3, 2) \rightarrow (1, 1) \rightarrow (2, 3) \rightarrow (0, 4) \rightarrow (2, 4)$

Shape: _____

Use with Grade 3, Chapter 8, Lesson 6, pages 172–173.

Interpret Line Graphs

Oscar read that crickets chirp more often as the temperature rises. He makes a line plot of the average number of times a cricket chirps at different temperatures.

Use the line graph for problems 1–9.

1. At 50° F, how many times does a cricket chirp per minute?

_____

2. At what temperature does a cricket chirp 80 times per minute?

_____

3. At 70° F, how many times does a cricket chirp per minute?

_____

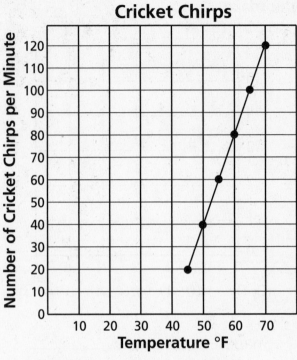

Cricket Chirps

4. At 45° F, how many times per minute does a cricket chirp?

_____

5. At what temperature does a cricket chirp 100 times per minute?

_____

6. At 55° F, how many times per minute does a cricket chirp?

_____

7. How is this line graph different from some other line graphs you have seen?

_____

8. What happens to the number of chirps per minute as the temperature drops?

_____

9. How is a line graph helpful?

_____

Use with Grade 3, Chapter 8, Lesson 7, pages 174–175.

Explore the Meaning of Multiplication

Suppose you are the leader of a marching band. Show six ways to arrange the 24 band members to make equal rows or groups. Do not use groups of 1 or 24. You may use models to help you.

When you increase the number of groups, what happens to the number in each group?

_____

_____

Relate Multiplication and Addition • Algebra

Draw lines to match the picture to the correct addition and multiplication sentences.

Addition Sentence		**Multiplication Sentence**

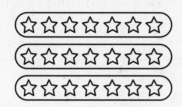

$4 + 4 + 4 + 4 + 4 = 20$ $5 \times 3 = 15$

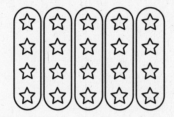

$4 + 4 + 4 + 4 = 16$ $3 \times 7 = 21$

$3 + 3 + 3 + 3 + 3 = 15$ $4 \times 4 = 16$

$7 + 7 + 7 = 21$ $5 \times 4 = 20$

How can you use addition to check a multiplication like 3×5?

_____

_____

_____

         Use with Grade 3, Chapter 9, Lesson 2, pages 192–194.

Explore Multiplication Using Arrays

Use the Commutative Property to write a different multiplication sentence. Then color an array on the stamp grid to match each multiplication sentence.

1. $3 \times 4 = 12$

_____

2. $2 \times 5 = 10$

_____

3. $6 \times 3 = 18$

_____

4. $1 \times 2 = 2$

_____

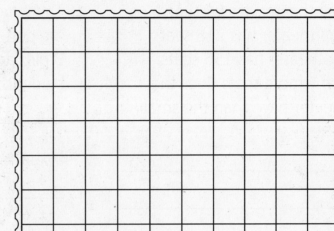
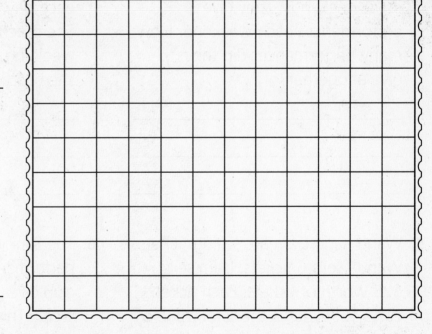

How are arrays 2×5 and 5×2 alike? How are the arrays different?

_____

_____

_____

Name _____

Problem Solving: Skill
Choose an Operation

Solve. Explain how you solved the problem.

1. The small theater has 84 seats. The large theater has 218 seats. Another theater has 156 seats. How many more seats does the large theater have than the small theater?

_____

_____

2. There are 4 groups of students who are putting on skits. Each group has 3 girls and 3 boys. How many students are putting on skits?

_____

_____

_____

3. The Cranford Players perform for 3 weekends. They give 3 performances each weekend. How many performances did they give altogether?

_____

_____

_____

4. Nell practices dance for 2 hours each day from Monday to Friday. She practices for 6 hours on the weekend. How many hours does she practice in 7 days?

_____

_____

_____

5. Wanda and Liam are selling tickets. Wanda sells 6 tickets. Liam sells $25 worth of tickets. Each ticket costs $5. What is the total value of the tickets that Wanda and Liam have sold?

_____

_____

_____

6. The Senior Acting Club gives 16 performances each season. The club gives 2 performances each week. This season it has already performed for 5 weeks. How many performances are left in the season?

_____

_____

_____

Multiply by 2 and 5

Find a multiplication message. Complete the problems. Then match the product in each box with a number in the message below. Put the letter from the box on the matching line. Read the message.

A	R	C
2 × 7 = _____	4 × 2 = _____	6 × 2 = _____
T	**F**	**O**
2 × 2 = _____	2 × 9 = _____	2 × 8 = _____

Message:

_____ _____ _____ _____ _____ _____
 18   14   12   4   16   8

P	U	E	L	R
2 × 5 = ____	5 × 4 = ____	5 × 5 = ____	7 × 5 = ____	5 × 6 = ____
L	**T**	**I**	**M**	**I**
5 × 7 = ____	3 × 5 = ____	9 × 5 = ____	5 × 8 = ____	5 × 9 = ____

Message:

_____ _____ _____ _____ _____ _____ _____ _____ _____ _____
 40   20   35   15   45   10   35   45   25   30

Look at the products that have 5 as a factor. What do you notice about the number in the ones place?

_____

_____

Name _____

Multiply by 3 and 4

Play this game with two to three players.
Take turns.
You will need a spinner.

How to Play

- Spin the spinner twice.

- Find the product of the
  two numbers.

- Enter the product from each round in the correct box.

- After Round 5, add all the products and enter the total.

The player with the highest total wins.

	Round 1	Round 2	Round 3	Round 4	Round 5	Total
Player 1						
Player 2						
Player 3						

How can you predict if the product of 3 and another number
will be even or odd?

_____

_____

_____

Name _____

Multiply by 0 and 1 • Algebra

Play the Go-or-No game with a partner. Take turns.

You will need two spinners.

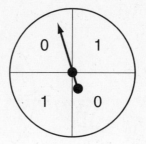 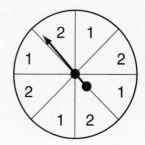

How to Play

• Spin both spinners.

• Multiply the two numbers to find the product. Move that number of spaces on the game board.

The player who crosses the finish line first is the winner.

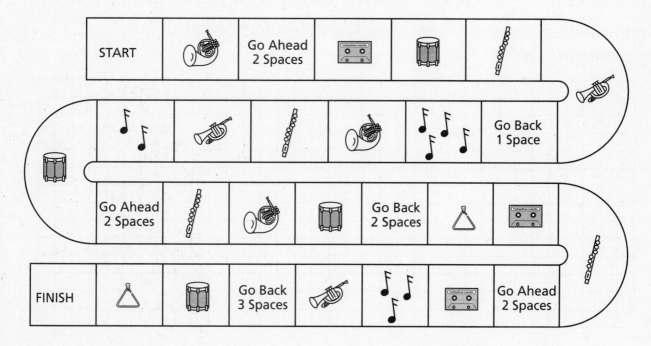

How can you tell the product of 1 × 0 × 4 without computing?

_____

_____

Name _____

Explore Square Numbers

Play "Square Off" with a partner. Take turns.
You will need crayons and a spinner.

How to Play

• Spin a number.

• Draw a square with that number of boxes on each side. Use the grid below. Color all the boxes in your square.

• You lose a turn if there is not enough room to draw your square.

• Play until all the boxes are colored. The player who colors more boxes wins.

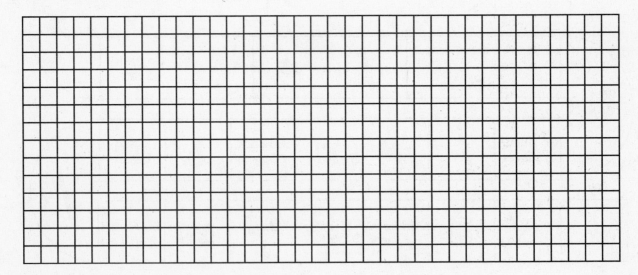

Suppose you spin a 9 on your first turn. Where would you draw your square? Explain your answer.

_____

_____

Use with Grade 3, Chapter 11, Lesson 1, pages 232–233.

Multiply by 6 and 8

Write a multiplication sentence. Tell how many legs for each group.

1. 6 ducks _____

2. 6 ants _____

3. 8 dogs _____

4. 8 boys _____

5. 6 horses _____

6. 8 birds _____

7. 6 spiders _____

8. 8 octopuses _____

9. 8 fish _____

10. 6 lions _____

Explain how you can use addition or multiplication to find the number of legs for a group of 8 ants.

_____

_____

Multiply by 7

Multiply the numbers in the inner rings. Write the product in the outer ring.

1.

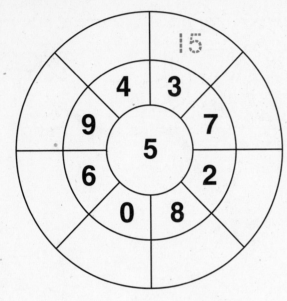

2.

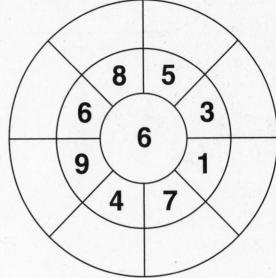

3.

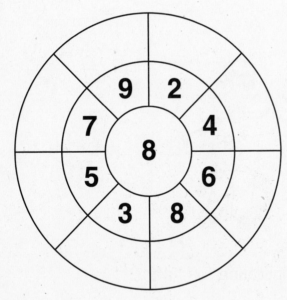

4.

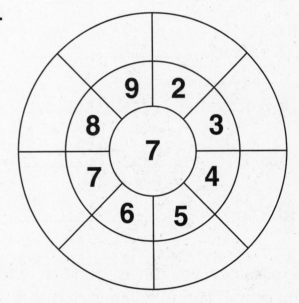

Tell how you find 7 × 8.

_____

_____

_____

Name _____

Find Missing Factors • Algebra

Find the missing factor. Match each number sentence to the picture that describes it. Write the number sentence in words.

1. 2 × ☐ = 12

   2 groups of ☐ is 12.

2. 4 ×  = 12

   _____

3. 1 × ☐ = 12

   _____

4. 3 × ☐ = 12

   _____

5. 6 × ☐ = 12

   _____

6. 12 × ☐ = 12

   _____

A.

B.

C.

D.

E.

F.

Problem Solving: Skill
Solve Multistep Problems

Write an additional question that helps you solve the problem.
Then solve the problem.

1. Two families go to Future World Park. There are 4 adults and
5 children. Children's tickets are $5 and adults' tickets are $8.
How much do the two families spend?

_____

_____

Solution: _____

2. Rick walked 1 mile to the train stop. Then he rode the train
5 miles to the museum. He returned home the same way.
How many miles did Rick travel on his trip from home to
the museum and back again?

_____

_____

Solution: _____

3. At the Empire State Building, Kate buys 2 T-shirts for
$8 each and 3 hats for $6 each. How much money
does Kate spend in all?

_____

_____

_____

Solution: _____

Use with Grade 3, Chapter 11, Lesson 5, pages 244–245.

Multiply by 10

Play "Lucky 10" with a partner. Take turns.
You will need two spinners.

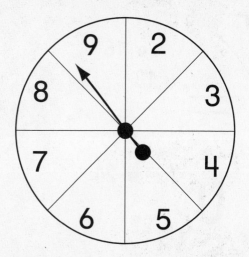

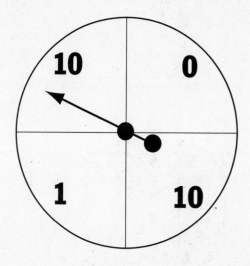

How to Play

• Spin both spinners.

• Multiply the two numbers. Write the product in the scoring chart.

• Play five rounds. Add the products from each round. The player with the greater number wins.

Scoring Chart						
Name	Round 1	Round 2	Round 3	Round 4	Round 5	Total

When you multiply a number by 10, what do you notice about the ones place of the product?

_____

_____

Multiply by 9

Complete the pattern.

Connect the dots in order for the products found when multiplying by 9.

$9 \times 1 =$ _____

$9 \times 2 =$ _____

$9 \times 3 =$ _____

$9 \times 4 =$ _____

$9 \times 5 =$ _____

$9 \times 6 =$ _____

$9 \times 7 =$ _____

$9 \times 8 =$ _____

$9 \times 9 =$ _____

$9 \times 10 =$ _____

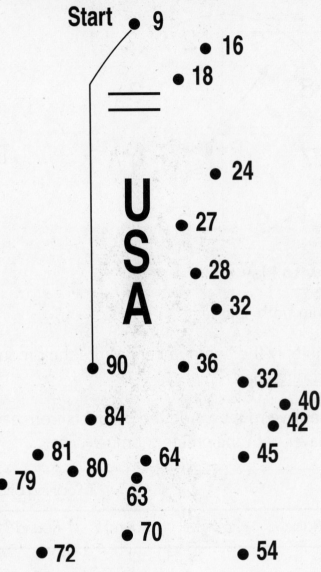

What do you notice about the sum of the digits in the products when you multiply each of the numbers 1 through 10 by 9?

_____

_____

Use with Grade 3, Chapter 12, Lesson 2, pages 252–253.

Multiplication Table

Use the strategies. Add on to a known fact or double a known fact to find a greater product.

X	0	1	2	3	4	5	6	7	8	9	10	11	12
0	0	0	0	0	0	0	0	0	0	0	0	0	0
1	0	1	2	3	4	5	6	7	8	9	10	11	12
2	0	2	4	6	8	10	12	14	16	18	20	22	24
3	0	3	6	9	12	15	18	21	24	27	30	33	36
4	0	4	8	12	16	20	24	28	32	36	40	44	48
5	0	5	10	15	20	25	30	35	40	45	50	55	60
6	0	6	12	18	24	30	36	42	48	54	60	66	72
7	0	7	14	21	28	35	42	49	56	63	70	77	84
8	0	8	16	24	32	40	48	56	64	72	80	88	96
9	0	9	18	27	36	45	54	63	72	81	90	99	108
10	0	10	20	30	40	50	60	70	80	90	100	110	120
11	0	11	22	33	44	55	66	77	88	99	110	121	132
12	0	12	24	36	48	60	72	84	96	108	120	132	144

Complete each way to find the product.

1. Find 14×8.

Use the 4s and 10s rows.

$14 \times 8 = (10 \times \boxed{}) + (\boxed{} \times 8)$

$\qquad = \boxed{} + \boxed{}$

$\qquad = \boxed{}$

2. Find 14×8.

Use the 7s row.

$14 \times 8 = (7 \times \boxed{}) + (\boxed{} \times 8)$

$\qquad = \boxed{} + \boxed{}$

$\qquad = \boxed{}$

Find each product.

3. $16 \times 5 =$ _____

4. $18 \times 7 =$ _____

5. $20 \times 11 =$ _____

Multiply 3 Numbers • Algebra

Play "Top Three" with a partner. Take turns.
You will need 30 index cards. Make five cards for each of
the numbers 0, 1, 2, 3, 4, 5.

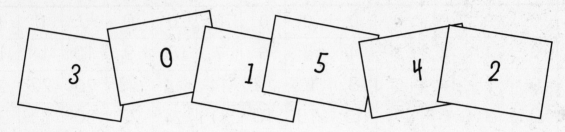

How to Play

- Shuffle the cards and place them in a pile facedown.

- Pick three cards.

- Find the product of the three numbers.

- The player with the greater product wins the round and
  collects all the cards for the round. Play until all the cards
  are used.

- The player with more cards wins.

$$4 \times 2 \times 5$$
$$= 4 \times 5 \times 2$$
$$= 4 \times 10$$
$$= 40$$

How can rearranging three factors help you find the product?

_____

_____

Use with Grade 3, Chapter 12, Lesson 4, pages 256–258.

Name _____

Explore the Meaning of Division

You can divide numbers of things in many different ways.
Show four ways to divide each number into groups of
equal size. Complete the table and the sentences. You
may use counters.

1. ☆ ☆ ☆ ☆ ☆ ☆ ☆ ☆ ☆ ☆ ☆ ☆   24 stars
   ☆ ☆ ☆ ☆ ☆ ☆ ☆ ☆ ☆ ☆ ☆ ☆

Number in All	Number of Groups	Number in Each Group	
24			____ groups of ____
24			____ groups of ____
24			____ groups of ____
24			____ groups of ____

2. 🚀🚀🚀🚀🚀🚀🚀🚀🚀🚀🚀🚀🚀🚀🚀🚀🚀🚀🚀🚀

20 rocket ships

Number in All	Number of Groups	Number in Each Group	
20			____ groups of ____
20			____ groups of ____
20			____ groups of ____
20			____ groups of ____

How does using counters help you divide things in many ways?

_____

_____

Relate Division and Subtraction • Algebra

Play a game in which you use repeated subtraction to divide.
Play with a partner. You'll need a spinner.

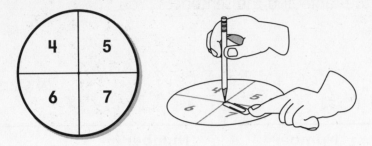

How to Play

- Make a spinner like the one shown above. Use a pencil and a paper clip to spin.

- Divide to complete the problems in the box below. Cut out the problems and place them face down on a tabletop.

- Take turns. Choose three problems from the group. Write each problem atop a sheet of paper as shown at right. The object of the game is to subtract the divisor (3 in the example) from the dividend (18) in each problem until you reach 0.

- Spin the spinner. When you land on the number you need, subtract. If you don't land on your divisor, the other player gets a turn to spin. Repeat until you complete the last problem.

The player who reaches 0 first wins the game.

Example

$$18 \div 3 = \underline{\hspace{1cm}}$$

$$
\begin{array}{r}
18 \\
- 3 \\
\hline
15 \\
- 3 \\
\hline
12 \\
- 3 \\
\hline
9 \\
- 3 \\
\hline
6 \\
- 3 \\
\hline
3 \\
- 3 \\
\hline
0
\end{array}
$$

Problems

35 ÷ 5 = _____	20 ÷ 5 = _____	24 ÷ 4 = _____
35 ÷ 7 = _____	20 ÷ 4 = _____	24 ÷ 6 = _____

Use with Grade 3, Chapter 13, Lesson 2, pages 280–282.

Name _____

Relate Multiplication to Division • Algebra

Complete. You will need two spinners.

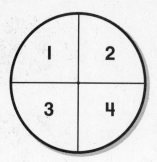 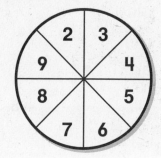 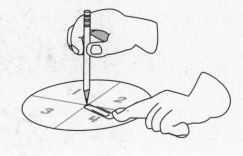

How to Play

• Take turns. Spin each spinner once.

• Player 1. Use the two numbers as factors to write a multiplication sentence. For example, if you spin a 4 and an 8, you could write $4 \times 8 = 32$.

• Player 2. Use the three numbers as the divisor, dividend, and quotient in a division sentence. For example, $32 \div 4 = 8$

• Repeat. Player 2 spins.

Record nine pairs of number sentences below.

1. _____	2. _____	3. _____
_____	_____	_____
4. _____	5. _____	6. _____
_____	_____	_____
7. _____	8. _____	9. _____
_____	_____	_____

Divide by 2

E 13-4 ENRICH

Start at 500. Skip-count backward by 2s. Connect the dots as you skip-count. Which planet do you see?

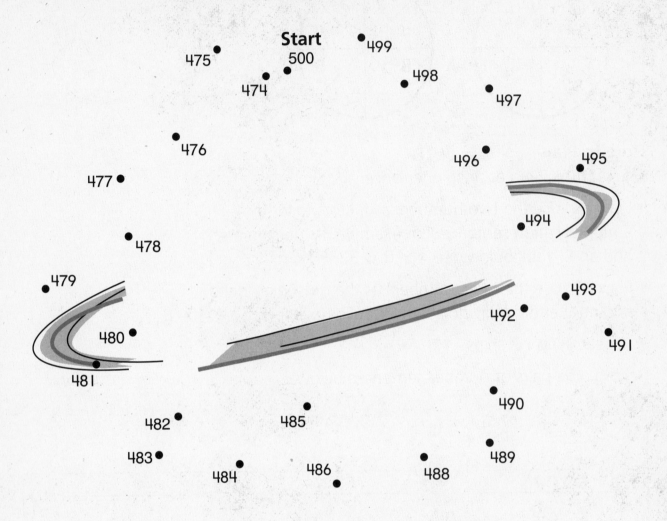

Describe a pattern you see when you skip-count by 2s.

_____

_____

Use with Grade 3, Chapter 13, Lesson 4, pages 288–289.

Problem Solving: Skill
Choose an Operation

Solve.

1. There are 20 students in science class. The students are in 5 equal groups. How many students are in each group?

_____

2. Denise is reading a chapter on the Sun. The chapter is 30 pages long. Denise has 6 pages left. How many pages has Denise read?

_____

3. Harry buys 27 star cards to give to 3 friends. He gives an equal number of cards to each friend. How many cards does each friend get?

_____

4. Virginia has 15 moon stickers and 5 sun stickers. How many stickers does she have in all?

_____

5. There are 5 students in each van going to the space museum. They are in 2 vans. How many students are going to the museum?

_____

6. Pauline has 40 star stickers. She uses 8 stickers and gives the rest of the stickers to her brother. How many stickers does Pauline give to her brother?

_____

7. Nell has 18 model rockets. She puts 6 rockets on one shelf. Nell puts the rest of the rockets on a second shelf. How many rockets does she put on the second shelf?

_____

8. Complete the tally chart. Show how many times you used each operation for problems 1–7.

Addition	
Subtraction	
Multiplication	
Division	

Divide by 5

You will need orange and yellow crayons or colored pencils.
Use yellow to color all the stars with numbers that can be divided evenly by 2.
Use orange to circle all the stars with numbers that can be divided evenly by 5.

List the numbers that can be divided evenly by 2.

_____

List the numbers that can be divided evenly by 5.

_____

Look at the numbers that can be divided evenly by 5. What pattern do you see?

_____

_____

Divide by 3

Blast off! Shade or color all the numbers that can be divided evenly by 3.

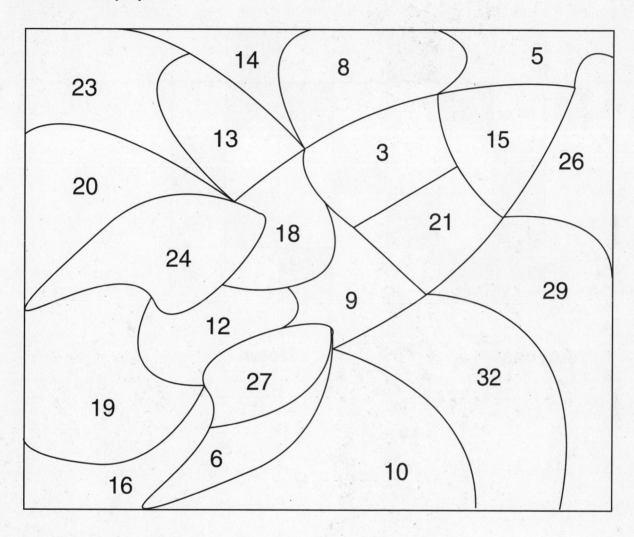

How can you use a product to find a related division fact?

_____

_____

_____

Divide by 4

Complete the crossword puzzle. Find each dividend.
Then use the clues to fill in the puzzle.

1.		▓	2.		▓	3.
	▓	4.		▓	▓	
▓	5.		▓	6.		▓
▓		▓	7.		▓	8.
▓	▓	9.		▓	10.	

Across

1. _____ ÷ 4 = 9

2. _____ ÷ 4 = 7

4. _____ ÷ 5 = 7

5. _____ ÷ 4 = 5

6. _____ ÷ 4 = 6

7. 80 ÷ 4 = _____

9. _____ ÷ 4 = 12

10. _____ ÷ 4 = 11

Down

1. _____ ÷ 4 = 8

2. _____ ÷ 5 = 5

3. _____ ÷ 3 = 9

4. _____ ÷ 5 = 6

5. _____ ÷ 6 = 4

6. _____ ÷ 5 = 4

7. _____ ÷ 7 = 4

8. _____ ÷ 3 = 8

Use with Grade 3, Chapter 14, Lesson 3, pages 300–302.

Divide with 0 and 1 • Algebra

Write +, −, ×, or ÷ in each circle to make each
number sentence true. Hint: Do the operation in parentheses first.

1. 5 $\bigcirc$ 5 = 1

2. 8 $\bigcirc$ 8 = 0

3. 6 $\bigcirc$ 6 = 12

4. 8 $\bigcirc$ 1 = 8

5. 8 $\bigcirc$ 1 = 9

6. 8 $\bigcirc$ 1 = 7

7. 3 ÷ (6 $\bigcirc$ 3) = 1

8. (4 $\bigcirc$ 4) ÷ 4 = 2

9. (6 $\bigcirc$ 6) ÷ 6 = 6

10. 12 ÷ (8 $\bigcirc$ 4) = 1

Write >, <, or = in each circle to make each
number sentence true. Hint: Do the operation in parentheses first.

11. 6 ÷ 6 $\bigcirc$ 6

12. 0 ÷ 7 $\bigcirc$ 0 + 7

13. 0 ÷ 4 $\bigcirc$ 0 × 4

14. 12 ÷ 2 $\bigcirc$ 3 + 3

15. 0 + 3 $\bigcirc$ 3 ÷ 3

16. 4 ÷ 1 $\bigcirc$ 1 + 4

17. (8 − 4) ÷ 4 $\bigcirc$ 0

18. 6 ÷ (6 ÷ 6) $\bigcirc$ 1

19. 0 ÷ 5 $\bigcirc$ 10 ÷ 10

20. 0 $\bigcirc$ (11 − 4) ÷ 7

What do you know about the dividend and the divisor
when the quotient is 1?

_____

_____

Divide by 6 and 7

Divide the number in the shaded part of the circle by the number in the center.

Write the answer in the outer part of the circle.

1.

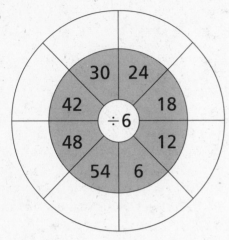

2.

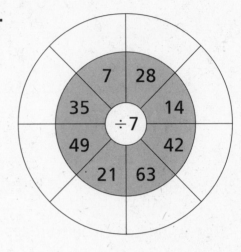

3.

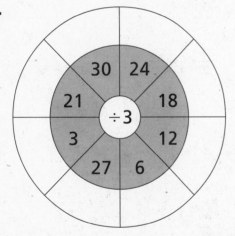

4.

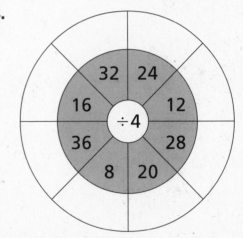

What do you notice when the same number is divided by 3 and by 6?

_____

_____

_____

Problem Solving: Skill
Solve Multistep Problems

Write the hidden question that helps you solve the problem.
Then solve the problem.

1. Matthew ties up 9 bags and James ties up 3 times as many
bags. They put an equal number of bags in 6 recycling bins.
How many bags do they put in each bin?

_____

_____

Solution: _____

2. Laurel collected 48 bottles. Hanna collected 6 fewer bottles.
They bagged the bottles in groups of 10 and took them to the
recycling center. How many bags were there in all?

_____

_____

Solution: _____

3. Tamala raises $6 by recycling bottles. She raises twice as much
by recycling cans. She plans to split the money equally between
2 charities. How much money will each charity receive?

_____

_____

Solution: _____

4. Danny collected 4 bags of cans per day for 5 days. Rich
collected half as many bags as Danny each day. At the end
of the fifth day, how many bags did the boys have?

_____

_____

Solution: _____

Divide by 8 and 9

Find the dividends to complete the cross-number puzzle.

Across

1. _____ ÷ 8 = 7

2. _____ ÷ 9 = 7

3. _____ ÷ 9 = 4

4. _____ ÷ 9 = 6

5. _____ ÷ 8 = 2

6. _____ ÷ 4 = 4

7. _____ ÷ 7 = 4

8. _____ ÷ 8 = 1

9. _____ ÷ 8 = 3

10. _____ ÷ 9 = 3

11. _____ ÷ 7 = 3

Down

1. _____ ÷ 9 = 6

2. _____ ÷ 8 = 8

3. _____ ÷ 9 = 4

4. _____ ÷ 8 = 7

5. _____ ÷ 9 = 2

6. _____ ÷ 7 = 2

7. _____ ÷ 9 = 3

8. _____ ÷ 9 = 9

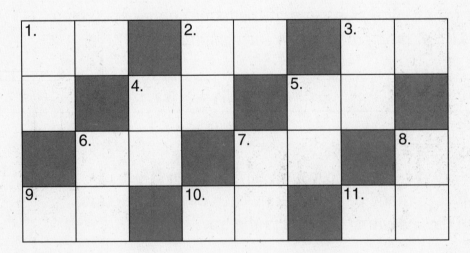

Use with Grade 3, Chapter 15, Lesson 3, pages 330–332.

Explore Dividing by 10

Mr. Olson surveyed students to find out which of the Great Lakes they have seen. Use the data in the table to complete the pictograph.

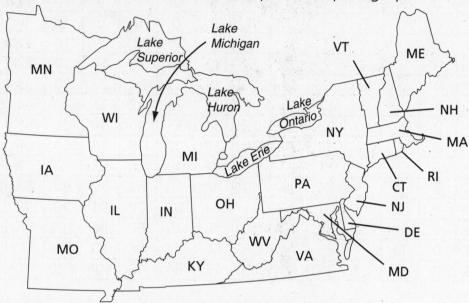

Great Lake	Number of Students
Erie	60
Huron	20
Michigan	90
Ontario	10
Superior	30

Great Lake	Number of Students
Erie	👤👤👤👤👤
Huron	👤👤
Michigan	
Ontario	
Superior	

Each 👤 stands for 10 students.

Use a Multiplication Table

Use the table to play a multiplication game.

×	0	1	2	3	4	5	6	7	8	9	10	11	12
0	0	0	0	0	0	0	0	0	0	0	0	0	0
1	0	1	2	3	4	5	6	7	8	9	10	11	12
2	0	2	4	6	8	10	12	14	16	18	20	22	24
3	0	3	6	9	12	15	18	21	24	27	30	33	36
4	0	4	8	12	16	20	24	28	32	36	40	44	48
5	0	5	10	15	20	25	30	35	40	45	50	55	60
6	0	6	12	18	24	30	36	42	48	54	60	66	72
7	0	7	14	21	28	35	42	49	56	63	70	77	84
8	0	8	16	24	32	40	48	56	64	72	80	88	96
9	0	9	18	27	36	45	54	63	72	81	90	99	108
10	0	10	20	30	40	50	60	70	80	90	100	110	120
11	0	11	22	33	44	55	66	77	88	99	110	121	132
12	0	12	24	36	48	60	72	84	96	108	120	132	144

How to Play

- Take turns with a partner.

- Write a multiplication or a division sentence. Give one of the numbers in your sentence to your partner.

- Your partner asks up to 10 questions to try to figure out your division sentence. For example, you say "36." Your partner may ask "Is 36 the product?" "Is one of the factors 6?"

- Record your multiplication or division sentences below. Use tally marks to keep track of the number of questions the other player asked. Play 5 rounds. Whoever asks fewer questions wins.

Multiplication or Division Sentence	Number of Questions Asked
1.	
2.	
3.	
4.	
5.	

Use with Grade 3, Chapter 15, Lesson 5, pages 336–337.

Use Related Facts • Algebra

Find each missing number.
Circle the fact that does not belong to each fact family.

1. $6 \times 7 =$ _____

$42 \div 6 =$ _____

$42 \div 7 =$ _____

$56 \div 7 =$ _____

2. $54 \div 9 =$ _____

$6 \times 9 =$ _____

$45 \div 5 =$ _____

$54 \div 6 =$ _____

3. $7 \times$ _____ $= 63$

$56 \div 7 =$ _____

$63 \div 9 =$ _____

_____ $\times 9 = 63$

4. $8 \div$ _____ $= 1$

$8 \times$ _____ $= 0$

$8 \div 1 =$ _____

$1 \times$ _____ $= 8$

5. _____ $\times 8 = 80$

$80 \div$ _____ $= 8$

$8 \times$ _____ $= 48$

$80 \div$ _____ $= 10$

6. $12 \times$ _____ $= 12$

_____ $\div 12 = 0$

$1 \times 12 =$ _____

_____ $\times 12 = 12$

7. $8 \times$ _____ $= 48$

_____ $\div 6 = 7$

$48 \div 8 =$ _____

$6 \times$ _____ $= 48$

8. _____ $\times 9 = 72$

$72 \div$ _____ $= 9$

$9 \times$ _____ $= 63$

$72 \div$ _____ $= 8$

9. $11 \times$ _____ $= 11$

_____ $\div 11 = 1$

$1 \times 10 =$ _____

_____ $\div 1 = 11$

How can you solve $56 \div n = 8$ using either a related
multiplication fact or a related division fact?

_____

_____

_____

Explore the Mean

Find the mean for each set of numbers. Match each set of numbers with the rearranged cubes to help find the mean.

1. 2, 4, 1, 1 _____

A.

2. 3, 2, 3, 4 _____

B.

3. 5, 1, 3 _____

C.

4. 6, 5, 1 _____

D.

5. 6, 5, 4 _____

E.

6. 5, 5, 2, 4 _____

F.

Use with Grade 3, Chapter 16, Lesson 3, pages 350–351.

Name _____

Find the Mean

Find the mean of each set of numbers. Then match each mean with a letter from the box. Write the letter above each line at the bottom of the page to find the name of a popular breed of cat.

A 2	C 7	E 3	I 9
M 4	N 6	O 5	T 8

1. 4, 5, 2, 5 _____

2. 3, 8, 9, 8, 10, 4, 7 _____

3. 12, 10, 11, 7, 5 _____

4. 5, 3, 8, 3, 10, 7 _____

5. 1, 2, 4, 5 _____

6. 8, 2, 6, 7, 9, 0, 3 _____

7. 8, 12, 14, 1, 5 _____

8. 7, 5, 4, 3, 6 _____

9. 8, 6, 3, 7, 6 _____

10. 1, 3, 4, 0 _____

11. 0, 2, 1, 5, 2 _____

12. 10, 8, 6, 4 _____

A popular breed of cat is the

____ ____ ____ ____ ____   ____ ____ ____ ____   ____ ____ ____
 1   10   3   9   5   12   8   6   4   2   11   7

Explore Multiplying Multiples of 10

Read the story. Follow directions. You will need crayons or markers, scissors, and paste.

Jill saves pairs of stickers in an album. Each pair has the same product. Color each pair a different color. Then cut out the stickers. Paste each pair in a section of the album.

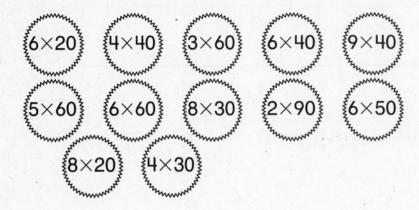

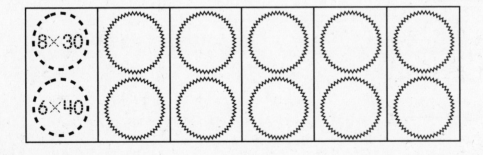

How did you find the product of 5 × 60?

_____

_____

Multiplication Patterns • Algebra

 17-2 ENRICH

Play Pinball Math! Write each product. Then trace the path your ball will follow. Try to hit 8 spots and get a total of 130,000 points before hitting the flippers.

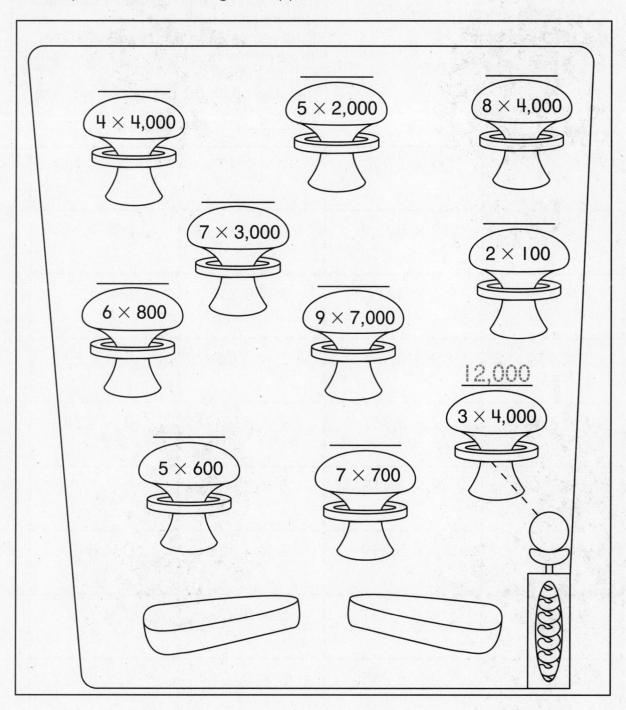

Estimate Products

Play a card game with a partner.
Cut out the cards along the lines.
Mix up the cards. Give each player
half the cards. Place the cards face
down on a tabletop.

How to Play

Each player turns over one card and
estimates the product by rounding.
The person with the **greater** estimated
product takes both cards.
If the estimated products are the same,
turn over the next card to see who wins
all four cards.
The player with the most cards wins the
game.

$3 \times 4,200$	$9 \times 4,326$	$7 \times 7,324$	$4 \times 8,248$
$5 \times 1,479$	8×84	2×586	$2 \times 7,314$
$2 \times 6,230$	4×421	$3 \times 9,316$	8×93
3×49	9×67	2×203	$6 \times 4,323$
$7 \times 3,345$	$2 \times 2,862$	7×674	$9 \times 1,249$
$6 \times 3,124$	$3 \times 3,115$	5×38	$4 \times 7,296$
9×738	$6 \times 5,298$	$8 \times 4,528$	6×648
$5 \times 4,716$	$4 \times 1,531$	5×896	3×729
4×311	$3 \times 8,245$	$8 \times 5,620$	$7 \times 6,222$

Use with Grade 3, Chapter 17, Lesson 3, pages 372–374.

Problem Solving: Skill

Find an Estimate or Exact Answer

Use the table to solve each problem. Explain why you used an estimate or an exact answer.

Seaside Gift Shop

Conch Shell	$3
T-shirt	$18
Beach Umbrella	$10
Bag of Shells	$7
Disposable Camera	$9

1. Jennie wants to buy 10 conch shells. She has two $20 bills. Does she have enough money?

_____

2. Alec buys 2 beach umbrellas and a bag of shells. He pays with three $10 bills. How much change does he receive?

_____

3. Mrs. Kaye buys 5 T-shirts for her family. About how much money does she spend?

_____

4. Andy buys 2 disposable cameras. He has a $10 bill and two $5 bills. Does he have enough money?

_____

5. There are about 15 large and 15 small shells in each bag. About how many shells are in 3 bags?

_____

Explore Multiplying 2-Digit Numbers

Use the clues to complete the crossmath puzzle.
Use place-value models to multiply.

Across

1. 6 × 34

3. 2 × 83

4. 4 × 16

5. 3 × 86

6. 5 × 29

8. 8 × 71

Down

2. 7 × 58

3. 3 × 47

5. 4 × 57

7. 9 × 62

Find the product of 6 × 34. Find the product of 6 × 30 and 6 × 4, and add them together. How is finding each answer the same? Different?

_____

_____

_____

_____

Name _____

Multiply 2-Digit Numbers

Write each product.

A 15 $\times 5$	**F** 97 $\times 4$	**A** 46 $\times 7$	**N** 58 $\times 2$	**S** 37 $\times 6$
75				
R 89 $\times 7$	**E** 46 $\times 4$	**B** 35 $\times 8$	**D** 64 $\times 3$	**E** 28 $\times 5$
U 72 $\times 8$	**K** 24 $\times 6$	**N** 42 $\times 2$	**O** 71 $\times 5$	**I** 33 $\times 9$
M 63 $\times 3$	**X** 23 $\times 9$	**Q** 98 $\times 2$	**C** 89 $\times 3$	**O** 55 $\times 4$

Use the products above to solve the riddles. Find the letter that goes with each answer. Write it on the line.

What do you call a carton of ducks?

A   __ __ __   __ __   __ __ __ __ __ __ __ __
75   280 355 207   220 388   196 576 322 267 144 140 623 222

If two's company and three's a crowd, what are four and five?

__ __ __ __
84 297 116 184

Name _____

Multiply Greater Numbers

E | 18-4
ENRICH

Write each product. Then circle each product in the number search puzzle below. Answers go horizontally, vertically, and diagonally. Then write the secret word that will be revealed in the number search puzzle.

1.	1,274	2.	2,345	3.	4,637	4.	9,727	5.	2,748
	× 3		× 3		× 5		× 8		× 7

6.	386	7.	58	8.	922	9.	224	10.	149
	× 4		× 2		× 9		× 4		× 6

11.	529	12.	128	13.	2,314	14.	5,362	15.	758
	× 9		× 3		× 4		× 5		× 3

16. $5 \times 115 = $ _____   **17.** $7 \times 9{,}322 = $ _____

18. $2 \times 219 = $ _____   **19.** $8 \times 837 = $ _____

20. $2 \times 248 = $ _____   **21.** $6 \times 453 = $ _____

3	8	2	2	0	6	5	1	9	2	2	7	4	3	7	0	3	5	5	2	7	1	8
8	1	5	0	9	5	4	9	8	6	9	4	3	9	7	2	0	4	2	3	1	3	9
4	7	6	1	2	2	4	2	1	8	2	9	8	6	8	9	6	3	9	1	5	4	4
5	3	0	1	7	5	2	3	1	1	5	8	0	3	1	8	1	7	3	8	4	7	8
9	2	5	6	4	4	9	6	5	0	2	9	5	1	6	6	9	6	0	5	1	6	5

The secret word is _____ .

Use with Grade 3, Chapter 18, Lesson 4, pages 390–391.

Choose a Computation Method

Use the numbers in each cloud to write the other factor for each multiplication problem.

1. (7, 1, 0)   6 × _____ = 642

2. (0, 2, 1)   3 × _____ = 360

3. (2, 5, 3)   _____ × 7 = 1,771

4. (0, 8, 4)   5 × _____ = 2,040

5. (8, 4, 3)   _____ × 3 = 1,449

6. (7, 1, 5)   2 × _____ = 1,502

7. (2, 6, 3)   _____ × 9 = 5,688

8. (0, 3, 1, 2)   4 × _____ = 4,128

Explore Dividing Multiples of 10

Write the quotients on the puzzle pieces. Cut out the pieces.
Arrange them in order from least to greatest quotients to get a
picture of an octagon.

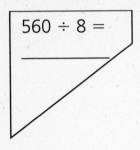

560 ÷ 8 =
_____

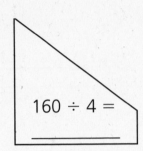

160 ÷ 4 =
_____

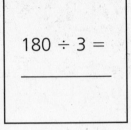

180 ÷ 3 =
_____

240 ÷ 8 =
_____

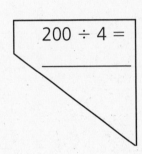

200 ÷ 4 =
_____

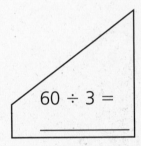

60 ÷ 3 =
_____

Think Critically

Why is it helpful to know your basic facts when you divide 360 ÷ 6?

_____

_____

_____

Use with Grade 3, Chapter 19, Lesson 1, pages 410–411.

Division Patterns • Algebra

Write the number that makes each sentence true. Color the boxes of all quotients less than 50 green. Color the boxes of all quotients from 50 through 90 yellow. Color the boxes of quotients greater than or equal to 100 blue.

320 ÷ 4 = **80**	210 ÷ 3 = _____	360 ÷ 4 = _____
420 ÷ 7 = _____	180 ÷ 2 = _____	400 ÷ 5 = _____
630 ÷ 9 = _____	640 ÷ 8 = _____	420 ÷ 6 = _____
480 ÷ 8 = _____	80 ÷ 4 = _____	540 ÷ 9 = _____
600 ÷ 2 = _____	180 ÷ 9 = _____	800 ÷ 8 = _____
600 ÷ 3 = _____	120 ÷ 3 = _____	400 ÷ 2 = _____
400 ÷ 4 = _____	200 ÷ 5 = _____	700 ÷ 7 = _____
800 ÷ 4 = _____	60 ÷ 2 = _____	900 ÷ 3 = _____

Think Critically

How are 80 ÷ 4 and 800 ÷ 4 the same and how are they different?

_____

_____

_____

Estimate Quotients

Find the hidden picture. Use compatible numbers. Write the
estimated quotients in the spaces. Color the estimated
quotients in the following way.

| 20 orange | 40 blue | 60 yellow | 30 red | 50 green | 70 purple |

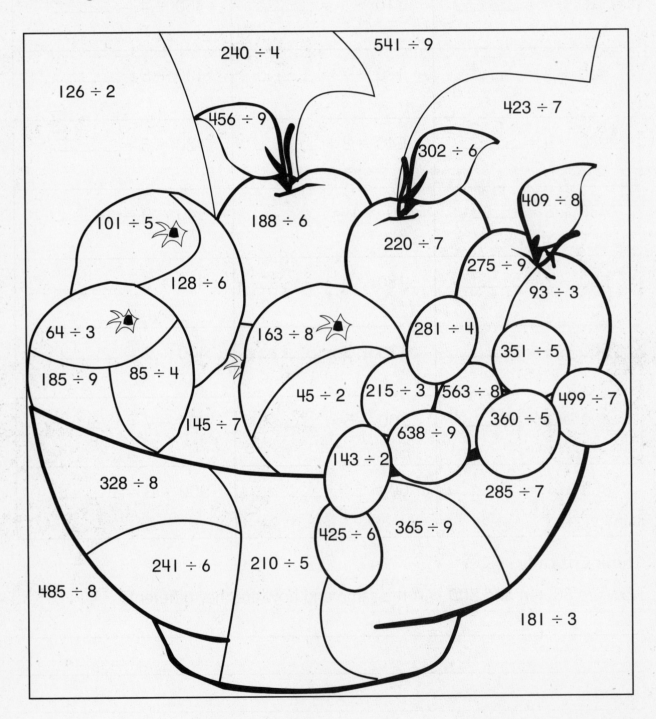

Use with Grade 3, Chapter 19, Lesson 3, pages 414–416.

Explore Division

Write each quotient.

T 5)70

T 3)93

S 6)90

L 2)38

B 4)96

U 2)40

R 6)96

A 9)90

E 3)84

C 7)91

T 5)90

U 4)68

P 3)78

I 8)88

U 3)75

P 7)84

Now solve each riddle. Write the letter that matches each quotient on the line.

What grows between your nose and your chin?

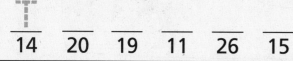

T
‾‾   ‾‾   ‾‾   ‾‾   ‾‾   ‾‾
14   20   19   11   26   15

What is a cup that you can't drink out of?

‾‾   ‾‾   ‾‾   ‾‾   ‾‾   ‾‾   ‾‾   ‾‾   ‾‾   ‾‾
10   24   25   31   18   28   16   13   17   12

Think Critically

When you divide a 2-digit number by 6, what is the greatest possible remainder you can have? How do you know?

_____

_____

_____

Divide 2-Digit Numbers

Help the farmer get to the greenmarket. Cut out the cards.
Mix them up and turn them facedown.
Take turns with a partner. You will need two markers.

How to Play

Pick a card. Find and check the quotient. If you are correct,
move your marker the number of spaces shown in the
circled number on the card. If you are wrong,
your partner gets two turns. Try to get to the
Farmers' Market first!

① $2\overline{)37}$	① $5\overline{)85}$	① $2\overline{)86}$	① $4\overline{)89}$	① $3\overline{)64}$
① $4\overline{)64}$	① $2\overline{)58}$	① $5\overline{)70}$	① $3\overline{)54}$	① $4\overline{)92}$
① $4\overline{)57}$	① $4\overline{)68}$	① $5\overline{)90}$	① $3\overline{)72}$	① $5\overline{)80}$
② $6\overline{)72}$	② $6\overline{)94}$	② $6\overline{)84}$	② $6\overline{)96}$	② $6\overline{)80}$
② $7\overline{)78}$	② $7\overline{)91}$	② $7\overline{)98}$	② $7\overline{)86}$	② $7\overline{)80}$
③ $8\overline{)90}$	③ $8\overline{)98}$	③ $8\overline{)92}$	③ $8\overline{)96}$	③ $9\overline{)99}$

Use with Grade 3, Chapter 19, Lesson 5 pages 420–422.

Problem Solving: Skill
Interpret the Remainder

Circle the correct word(s) or number(s) to make each statement true.

1. There are 46 people at a ceremony at Town Hall. They sit at tables that have 8 seats each.

   There are   8   6 or 8   people at each table.

   There are   5   6   tables used.

   Explain your thinking: _____

2. The Town Historical Society sells raffle tickets for $6. Ms. Winston has $20.

   Ms. Winston can buy   3 $3\frac{1}{2}$ 4   raffle tickets.

   If Ms. Winston buys the greatest possible number of tickets, she will have   less than $1   more than $1   left.

   Explain your thinking: _____

3. There are 35 people who do volunteer work in a park in the next town. They drive to the park in vans that can hold 6 people.

   There are   5   5 or 6   6 or 7   people in each van.

   They use   5   6   vans.

   Explain your thinking: _____

4. In 2003, the walk-a-thon set a record by collecting $900. The amount was split equally between two local charities.

   The 2003 amount was   more than   less than   the amount collected in 2002.

   Each charity received $400   more than $400   less than $400.

   Explain your thinking: _____

Divide 3-Digit Numbers

Take a walk in a garden.

START

$336 \div 4 =$ _____ $\div 3 =$ _____ $\div 2 =$ _____

START

$432 \div 2 =$ _____ $\div 6 =$ _____ $\div 18 =$ _____

START

$990 \div 3 =$ _____ $\div 11 =$ _____ $\div 5 =$ _____

Quotients with Zeros

Choose a number from 2–9 as a divisor for each division problem
below. If your quotient has a zero in it, you get 2 points (a remainder
of zero doesn't count). If your quotient doesn't have a zero, you get
only 1 point. The highest possible score is 20. You can use a divisor more
than once.

1. 104
6)624

2. __)812

3. __)914

4. __)636

5. __)864

6. __)211

7. __)535

8. __)723

9. __)918

10. __)836

Make up your own division exercises that have the following quotients.

11. 202   )‾‾‾

12. 109   )‾‾‾

13. 102   )‾‾‾

14. 308   )‾‾‾

How did you decide what the divisor and dividend should be in problem 11?

_____

_____

_____

Choose a Computation Method

One of the largest birds can fly for hours, sometimes even days, without flapping its wings. What is the name of this bird? To find out, find each quotient and write the matching letter in each space below.

1. 3)‾421‾ B	**2.** 9)‾790‾ S	**3.** 6)‾378‾ O
4. 6)‾1,948‾ A	**5.** 8)‾189‾ R	**6.** 5)‾2,381‾ L
7. 4)‾932‾ A	**8.** 2)‾589‾ S	**9.** 7)‾978‾ T

The name of the bird is

___   ___   ___   ___   ___   ___   ___   ___   ___

233   476 R1   140 R1   324 R4   139 R5   23 R5   63   294 R1   87 R7

Use with Grade 3, Chapter 20, Lesson 3, pages 436–437.

Explore Lengths

Estimate the length of each piece of ribbon. Then use small paper clips and an inch ruler to measure to the nearest inch.

1.

Estimate with paper clips _____   Estimate in inches _____

Length in paper clips _____   Length in inches _____

_____

2.

Estimate with paper clips _____   Estimate in inches _____

Length in paper clips _____   Length in inches _____

_____

3.

Estimate with paper clips _____   Estimate in inches _____

Length in paper clips _____   Length in inches _____

_____

4.

Estimate with paper clips _____   Estimate in inches _____

Length in paper clips _____   Length in inches _____

Explore Customary Units of Length

You will need an inch ruler to measure. Follow each direction.
Write the answer to each question on the line.

1. From START, draw a line to the right exactly 6 inches long.
   Write the letter where you are now. _____

2. From the point from problem 1, draw a line down 2 inches.
   Write the letter where you are now. _____

3. Draw a line from the point where you are now to point *J*.
   Write the length of the line you just drew. _____

4. Draw a line exactly 4 inches long from *J* to point *K, L, M,* or *N*. Write the
   name of the point where you are now. _____

5. Draw a line from the point where you are now to START. Write the length
   of the line you just drew. _____

6. You just drew an important place on a baseball field. What did you draw?

   _____

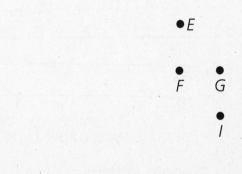

Use with Grade 3, Chapter 21, Lesson 2, pages 456–457.

Customary Units of Capacity

Riddle: Why is basketball such a messy sport?

Find the answer to the riddle.

Circle the estimated capacity for each item. Then write the letters you circled in order on the blank lines below.

1. a mug

    Y 1 c   **Z** 1 pt

2. a sink

    N 1 c   **O** 1 gal

3. a very large pot

    T 5 pt   **U** 5 gal

4. a tea kettle

    D 2 qt   **E** 2 gal

5. a juice box

    R 1 c   **S** 1 qt

6. a lunch box

    I 1 qt   **J** 1 gal

7. a beach pail

    B 20 c   **C** 20 gal

8. a fish tank

    A 2 qt   **B** 2 gal

9. a trash barrel

    K 20 c   **L** 20 gal

10. a squirt gun

    E 1 c   **F** 1 qt

11. a dog's dish

    A 1 pt   **B** 1 gal

12. an ice chest

    K 5 c   **L** 5 gal

13. a sandwich bag

    L 2 c   **M** 2 qt

14. a bathtub

    N 20 pt   **O** 20 gal

15. a cake pan

    V 2 pt   **W** 2 gal

16. a mixing bowl

    D 1 c   **E** 1 gal

17. a soup bowl

    R 2 c   **S** 2 qt

___ ___ ___   ___ ___ ___ ___

___ ___ ___   ___ ___ ___ ___

Customary Units of Weight

Cut out the puzzle pieces at the bottom of the page. Match each item with its estimated weight. Place the pieces in the same order as shown below. Write the name of the sports item you made.

about 1 oz	about 5 oz	about 180 lb	about 12 oz
about 2,500 lb	about 1 lb	about 5 lb	about 60 lb

This is a _____.

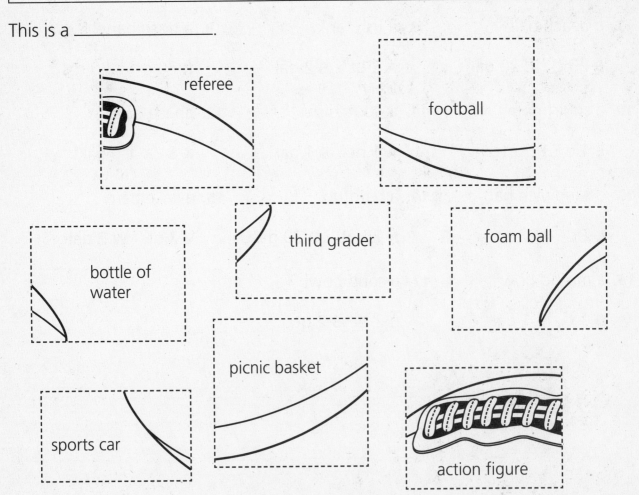

referee

football

bottle of water

third grader

foam ball

picnic basket

sports car

action figure

Use with Grade 3, Chapter 21, Lesson 4, pages 460–461.

Convert Customary Units

Play this game of concentration with a partner. Cut out the cards below and mix them up. Turn the cards face down and spread them out.

To play: Take turns. One player turns over two cards. If they show equal lengths, capacities, or weights, they match. Players keep the matches they make.

How can you tell that 10 cups equals 2 quarts 1 pint?

_____

_____

_____

2 cups	1 pint	12 inches	1 foot
4 quarts	1 gallon	3 feet	1 yard
2 pints	1 quart	16 ounces	1 pound
2 pints	4 cups	8 pints	1 gallon
36 inches	1 yard	6 feet	2 yards
24 inches	2 feet	18 inches	1 foot 6 inches
4 pints	2 quarts	10 cups	2 quarts 1 pint
32 ounces	2 pounds	20 ounces	1 pound 4 ounces

Name _____

Problem Solving: Skill
Check for Reasonableness

Circle the statement that is reasonable.

1. Melanie jumps 40 inches. Sandy jumps 1 yard.

Melanie says, "I jumped farther than Sandy."

Sandy says, "I jumped the same distance as Melanie."

Explain your thinking: _____

_____

2. Chen throws the shot-put ball 98 inches. Pat throws the shot-put ball 8 feet 2 inches.

Chen says, "I threw the ball just as far as Pat did."

Pat says, "I threw farther than Chen."

Explain your thinking: _____

_____

3. Robert kicks a kickball 20 yards. Ursula kicks it 62 feet. Jane kicks it 58 feet.

Jane says, "Robert kicked the shortest distance."

Ursula says, "I kicked the longest distance."

Explain your thinking: _____

_____

4. Tanya throws a ball 40 yards. Sam throws a ball 110 feet.

Tanya says, "I threw the ball twice as far as Sam did."

Sam says, "Tanya threw the ball only 10 feet farther than I did."

Explain your thinking: _____

_____

Use with Grade 3, Chapter 21, Lesson 6, pages 466–467.

Explore Metric Units of Length

Complete. You will need a ruler that shows centimeters.

Using the metric ruler, measure each soccer kick.
Write the number of centimeters on each line.

1.

4 cm _____   _____   _____

2.

_____   _____   _____

3.

_____   _____

Solve. Follow directions.

1. Underline the soccer player who kicked the ball 7 cm.

2. Put a square around the soccer ball that went a total of 11 cm.

3. Put a circle around the two soccer players who kicked 24 cm in all.

4. One soccer ball went into the goal after a 3-cm kick. Circle that goal.

Metric Units of Capacity

Find two items for each capacity. Draw a line from each item to the capacity.

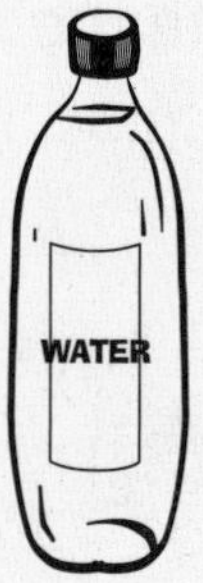

1L

5L

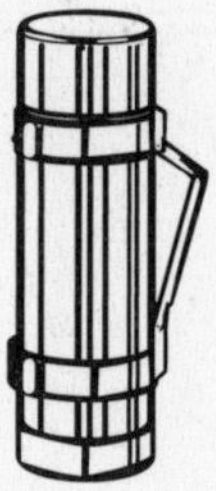

1 mL

250 mL

Is 250 mL nearly a liter? Explain your choice.

Use with Grade 3, Chapter 22, Lesson 2, pages 474–475.

Name _____

Metric Units of Mass

To solve these problems, you will need a balance scale, some gram masses, and sheets of paper.

Use the balance scale to find the mass of some items. Use patterns to help you solve the problems.

1. Find the mass of 1 sheet of paper.   _____

2. What is the mass of 5 sheets of paper?   _____

3. What is the mass of 50 sheets of paper?   _____

4. Paper is usually sold in reams of 500 sheets.
How many kilograms is the mass of 1 ream
of paper? Explain how you got your answer.

_____

_____

_____

5. A carton of paper usually holds 10 reams of paper.
What is the mass of a carton of paper?   _____

6. A metric ton is equal to 1,000 kg.
Would you need 5, 50, 500, or 5,000 cartons of
paper to get to a mass of a metric ton?
Explain how you got your answer.

_____

_____

Convert Metric Units

Look down each column. Find the greatest length, capacity, or mass.
Color that space yellow.

1.

1 m	10 cm	15 cm	75 cm	120 cm	115 cm	20 cm
1 dm	2 dm	40 cm	1 m	1 m	15 dm	2 dm
1 cm	1 dm	5 dm	9 dm	11 dm	1m 5cm	2m

2.

1 mL	500 mL	1,000 mL	1,500 mL	2 L 50mL	2,500 mL	4 L 600mL
100 mL	1 L	1 L	2 L	2,000 mL	3,000 mL	4,000 mL
1 L	800 mL	1,200 mL	1 L	2 L 500mL	2 L	5 L

3.

500 g	500 g	700 g	1 kg 800 g	2 g	300 g	3,500 g
50 g	1 kg	1 kg	1,500 g	2 kg	2 kg	3 kg
5 g	50 g	1,500 g	1kg	200 g	3,000 g	2 kg 900 g

4. If you want to change from larger units to smaller units,
do you multiply or divide? How do you know?

_____

_____

_____

Temperature

Use the clues to find each temperature. Color the
thermometers to show each temperature.

What's the temperature?

1. In Baltimore, Maryland, the average temperature
   in January is the freezing point of water.   _____ °F

2. In Phoenix, Arizona, in August the temperature
   is 8 less than 10 × 10.   _____ °F

3. In Houston, Texas, in January the temperature
   is 5 tens.   _____ °F

4. In Fresno, California, in August the temperature
   is 3 less than 6 × 5.   _____ °C

5. In Portland, Oregon, in January the temperature
   is 2 × 2 × 10.   _____ °F

6. In Asheville, North Carolina, in August the temperature
   is an even number less than 40 that has the same
   number in the tens and ones place.   _____ °C

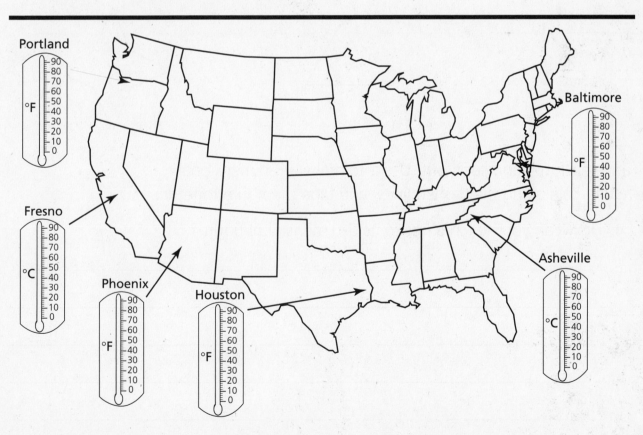

3-Dimensional Figures

Which net can be folded to make the 3-dimensional figure shown? Circle it.

1.

 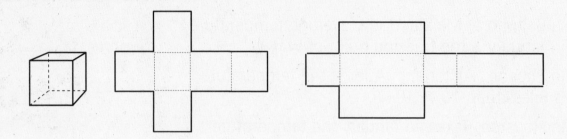

2.

 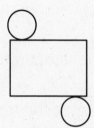 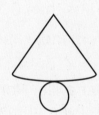

3.

 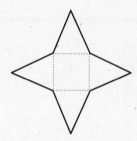

Trace each net and cut it out. Use it to make a 3-dimensional figure. Fold along the dashed lines, and tape the figure together.

4. How can you match a net to its 3-dimensional figure?

_____

_____

_____

_____

2-Dimensional Figures

Look at the puzzle. Find as many circles, triangles, squares, and rectangles as you can. Then answer the questions.

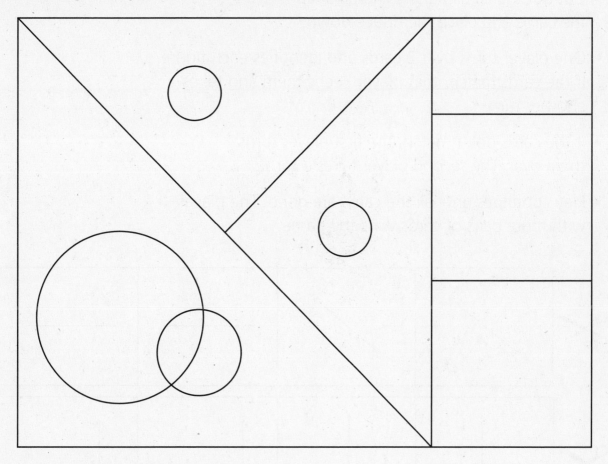

1. How many circles do you see? _____

2. How many triangles do you see? _____

3. How many squares do you see? _____

4. How many rectangles do you see? _____

5. Which 3-dimensional figure could you use to trace each figure in the puzzle?

_____

_____

_____

Lines, Line Segments, Rays, and Angles

Play a memory game with a partner. Take turns.

- Cut out the cards below. Then mix up the cards and place them face down.

- One player turns over 2 cards and identifies each figure. If the cards match, that player keeps them and takes another turn.

- If the cards don't match, the first player turns them over. The second player takes a turn.

- Play continues until all the cards are gone. The player with more pairs of cards wins the game.

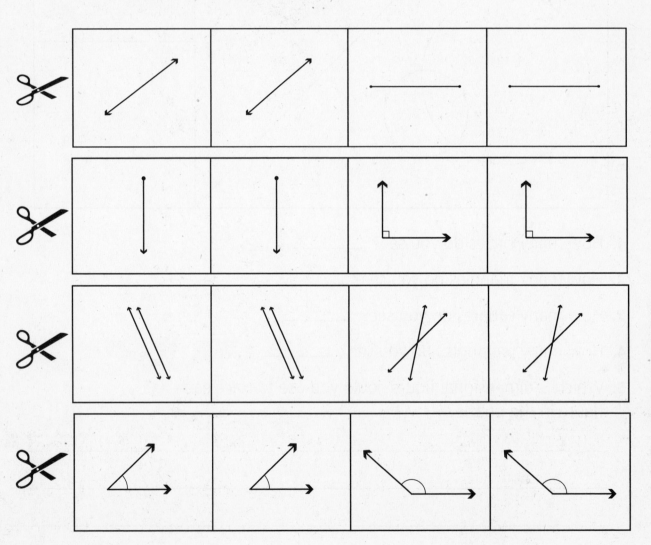

Use with Grade 3, Chapter 23, Lesson 3, pages 508–510.

Polygons

Trace each polygon and cut it out. Identify the polygon you could make by putting the polygons together.

1.

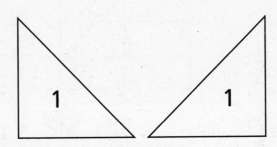

_____

2.

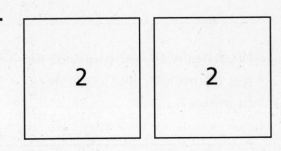

_____

3.

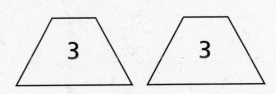

_____

4.

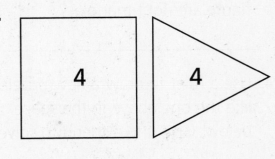

_____

5.

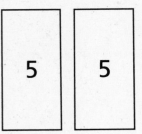

_____

6.

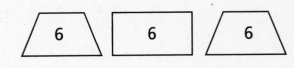

_____

7. How could you change the polygon you made in puzzle 6 to a hexagon?

_____

_____

_____

Triangles

Complete.

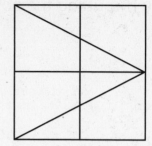

1. How many triangles are in the figure? Hint: There are more than 8.

_____

2. How many of the triangles in the figure look like equilateral triangles?

_____

3. How many of the triangles in the figure are right triangles?

_____

4. Use a ruler to draw a new puzzle like the one above in the space below. Give it to a friend to solve.

5. Draw a right scalene triangle.

Quadrilaterals

Identify the figure that does not belong. Give a reason.

1.

_____

_____

2.

_____

_____

3.

_____

_____

4.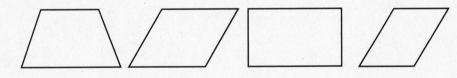

_____

_____

5.

_____

_____

Problem Solving: Skill
Use a Diagram

Use the illustration to solve problems 1 and 2.

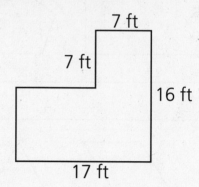

1. Nina designed this room. What two shapes make up this room?

_____

2. How many labels are missing from this diagram? How long is each unlabeled side?

_____

3. Draw a design of a room below. Label each side.

_____

4. What shapes make up your room?

_____

5. Write a problem using your diagram. Ask a friend to solve it.

_____

Congruent and Similar Figures

This pattern is made from congruent triangles.

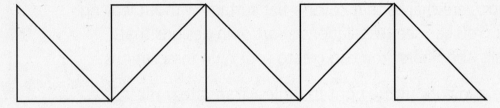

This pattern is made from similar triangles.

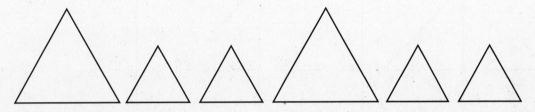

1. Choose a shape. Create a pattern using congruent shapes.
   Draw your pattern below.

2. Choose another shape. Create a pattern using similar shapes.
   Draw your pattern below.

3. Create a third pattern using shapes that are not congruent or similar.
   Draw your pattern below.

4. Are all triangles similar? Explain.

   _____

   _____

Name_____

Explore Translations, Reflections, and Rotations

Tessellations are shapes that cover a flat surface without leaving any gaps. You can see tessellations in art or in designs that cover walls and floors. You can create your own tessellations.

1. Trace and cut out the right triangles. Translate, reflect, and rotate the triangles to tessellate the square shown.

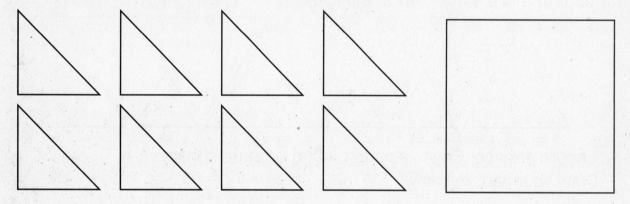

2. Trace and cut out the hexagons. Translate, reflect, and rotate the hexagons to tessellate the figure shown.

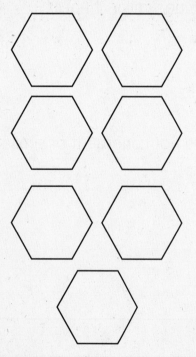
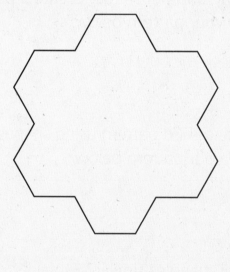

Use with Grade 3, Chapter 24, Lesson 2, pages 526–527.

Explore Symmetry

Complete each shape to make it symmetrical.

1.

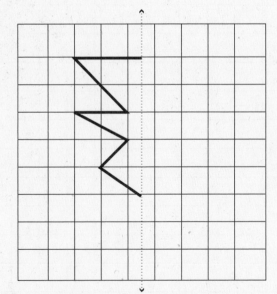

2.

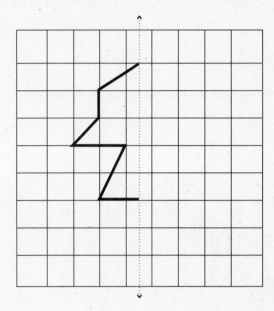

3.

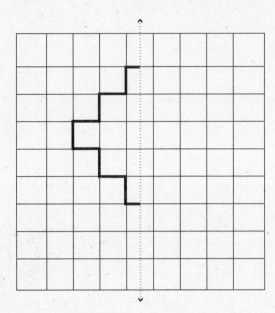

4.

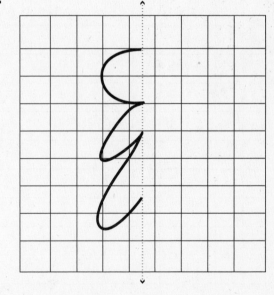

5. How can you prove to another student that each shape is symmetrical?

_____

_____

_____

Perimeter

How many shapes can you draw with a perimeter of exactly 12 units? Draw them below. If you like, decorate your shapes.

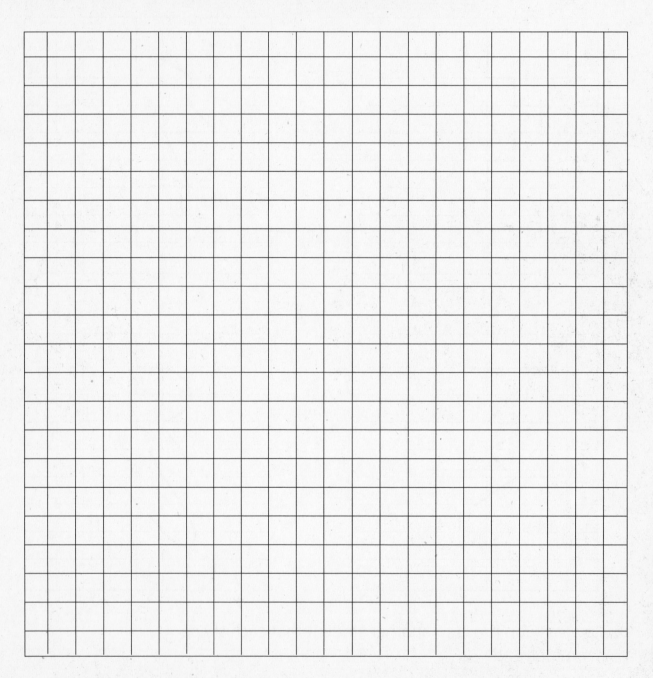

Does a given perimeter mean only one shape can be formed? Explain.

_____

_____

Use with Grade 3, Chapter 24, Lesson 5, pages 532–533.

Area

1. How many rectangles can you draw with an area of exactly 12 square units? Draw them below.

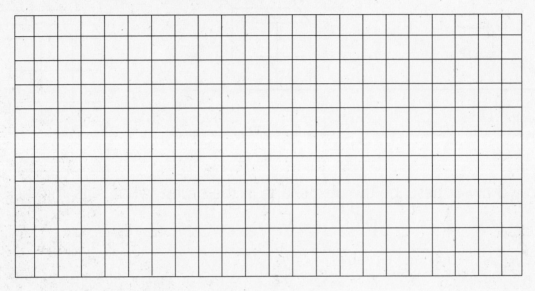

2. How many squares can you draw with an area of 25 square units or less? Draw them below.

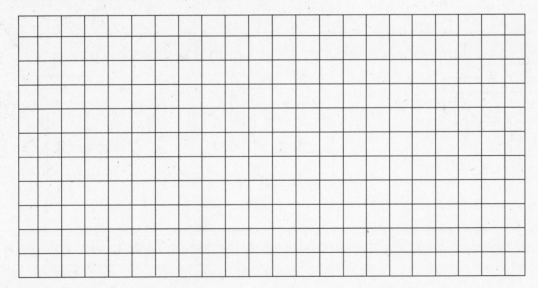

3. Do figures with the same area always have the same perimeter? Explain.

_____

_____

Explore Volume

1. Find the volume of each cube in the series in cubic units.

  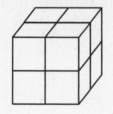 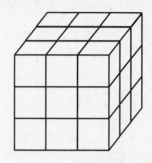

_____   _____   _____

2. Now draw or describe the next shape in the series. What is its volume?

_____

3. Find the volume of each rectangular prism in the series in cubic units.

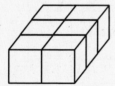 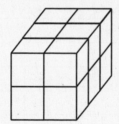 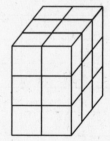

_____   _____   _____

4. Now draw or describe the next shape in the series. What is its volume?

_____

Use with Grade 3, Chapter 24, Lesson 7, pages 538–539.

Parts of a Whole

Figure out the pattern. Shade the last figure to show the next step in the pattern.

1.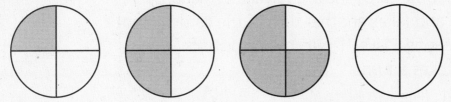

2.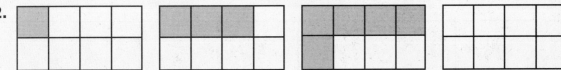

3.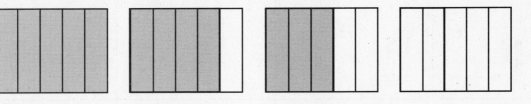

4.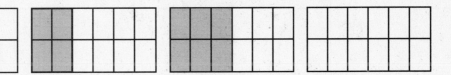

5.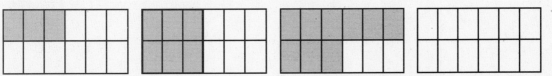

6. A square is divided into 8 parts. All 8 parts are shaded as shown. What fraction could you use to show the fraction of the square that is shaded?

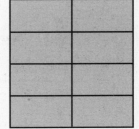

_____

Explore Equivalent Fractions

Play a math game with a partner. Cut out the cards below. Mix up the cards and place them facedown in a 4-by-6 array. Player 1 turns over 2 cards. If the cards show equivalent fractions, Player 1 keeps them and takes another turn. If the cards do not show equivalent fractions, Player 1 puts the cards back facedown. Player 2 takes a turn.

The player with more pairs of equivalent fractions wins.

$\frac{1}{2}$	$\frac{1}{2}$	$\frac{2}{4}$	$\frac{2}{4}$
$\frac{1}{3}$	$\frac{1}{3}$	$\frac{2}{6}$	$\frac{2}{6}$
$\frac{3}{4}$	$\frac{3}{4}$	$\frac{6}{8}$	$\frac{6}{8}$
$\frac{1}{5}$	$\frac{1}{5}$	$\frac{2}{10}$	$\frac{2}{10}$
$\frac{3}{6}$	$\frac{3}{6}$	$\frac{4}{8}$	$\frac{4}{8}$
$\frac{2}{3}$	$\frac{2}{3}$	$\frac{4}{6}$	$\frac{4}{6}$

Use with Grade 3, Chapter 25, Lesson 2, pages 558–559.

Fractions in Simplest Form

Circle all the fractions that are in simplest form.

D	N	C	K	E	T
$\frac{8}{12}$	$\frac{4}{12}$	$\frac{1}{4}$	$\frac{2}{4}$	$\frac{1}{5}$	$\frac{3}{9}$

G	S	O	J	Z	A
$\frac{5}{10}$	$\frac{4}{8}$	$\frac{4}{16}$	$\frac{2}{6}$	$\frac{3}{12}$	$\frac{6}{8}$

M	B	R	X	Q	R
$\frac{5}{6}$	$\frac{3}{12}$	$\frac{2}{5}$	$\frac{10}{12}$	$\frac{3}{6}$	$\frac{1}{6}$

Y	L	U	V	W	P
$\frac{4}{9}$	$\frac{5}{15}$	$\frac{5}{6}$	$\frac{2}{8}$	$\frac{4}{10}$	$\frac{8}{10}$

Write the letters that are above each circled fraction.

___ ___ ___ ___ ___ ___ ___

Unscramble the letters to answer the question.

Which planet is covered with craters that are named after famous writers, artists, and composers?

___ ___ ___ ___ ___ ___ ___

Compare and Order Fractions

Be a fraction detective. Use the clues to find each fraction.

1. I'm a fraction greater than $\frac{1}{2}$ and less than $\frac{3}{4}$. My numerator is 5 times 1. What fraction am I?

_____

2. I'm a fraction equivalent to $\frac{6}{8}$. My denominator is 2 times 2. What fraction am I?

_____

3. I'm a fraction less than $\frac{1}{4}$. My denominator is 2 times 4. What fraction am I?

_____

4. I'm a fraction less than $\frac{1}{2}$, but greater than $\frac{1}{3}$. My denominator is 3 times 4. What fraction am I?

_____

5. I'm a fraction equivalent to $\frac{1}{2}$. My numerator is 2 times 2. What fraction am I?

_____

6. You can find me on a number line between $\frac{1}{2}$ and $\frac{3}{4}$. My numerator is 2. What fraction am I?

_____

7. I'm a fraction equivalent to $\frac{1}{3}$. My numerator is 2. What fraction am I?

_____

8. I'm a fraction with a numerator of 3. My denominator is 2 times my numerator. What fraction am I?

_____

9. I'm a fraction greater than $\frac{3}{4}$ and less than 1. To show me, shade 7 parts of the whole. What fraction am I?

_____

10. You can find me on a number line between $\frac{1}{4}$ and $\frac{1}{6}$. My numerator is 1. What fraction am I?

_____

Parts of a Group

Word	Word Fraction	Fraction
math	$\frac{1\ \text{vowel}}{4\ \text{letters}}$	$\frac{1}{4}$
gardens	$\frac{2\ \text{vowels}}{7\ \text{letters}}$	$\frac{2}{7}$

Write word fractions for each word in the table below.
Then write the fraction.

	Word	Word Fraction	Fraction
1.	hello	_____	_____
2.	teacher	_____	_____
3.	friend	_____	_____
4.	cat	_____	_____
5.	name	_____	_____
6.	book	_____	_____
7.	mom	_____	_____
8.	girl	_____	_____
9.	boy	_____	_____
10.	goodbye	_____	_____

11. If $\frac{2}{5}$ of the letters in a word are vowels, what fraction of the letters in the word are consonants? Give an example.

_____

Explore Finding Parts of a Group

Skip-count. Look for a pattern. Use it to find each fraction of a number.

1. $\frac{1}{6}$ of 12 = 2

$\frac{2}{6}$ of 12 = 4

$\frac{3}{6}$ of 12 = 6

$\frac{4}{6}$ of 12 = 8

$\frac{5}{6}$ of 12 = _____

$\frac{6}{6}$ of 12 = _____

2. $\frac{1}{5}$ of 15 = 3

$\frac{2}{5}$ of 15 = 6

$\frac{3}{5}$ of 15 = 9

$\frac{4}{5}$ of 15 = _____

$\frac{5}{5}$ of 15 = _____

3. $\frac{1}{4}$ of 40 = 10

$\frac{2}{4}$ of 40 = 20

$\frac{3}{4}$ of 40 = _____

$\frac{4}{4}$ of 40 = _____

4. $\frac{1}{6}$ of 24 = 4

$\frac{2}{6}$ of 24 = 8

$\frac{3}{6}$ of 24 = _____

$\frac{4}{6}$ of 24 = _____

$\frac{5}{6}$ of 24 = _____

$\frac{6}{6}$ of 24 = _____

5. $\frac{1}{5}$ of 30 = 6

$\frac{2}{5}$ of 30 = 12

$\frac{3}{5}$ of 30 = _____

$\frac{4}{5}$ of 30 = _____

$\frac{5}{5}$ of 30 = _____

Use with Grade 3, Chapter 25, Lesson 6, pages 568–569.

Problem Solving: Skill
Check for Reasonableness

Circle the statement that helps you solve the problem.
Then solve the problem.

1. Sarah bought a 1-pound bag of flour. She has $\frac{1}{2}$ bag left.
How many ounces of flour does Sarah have left?

There are 16 ounces in 1 pound.

A pound is heavier than an ounce.

Solution: _____

2. Mark makes 2 dozen brownies at home. He brings $\frac{3}{4}$ of the
brownies to his class. How many brownies does he have left
at home?

$\frac{3}{4}$ is less than 2 dozen.

2 dozen means 24.

Solution: _____

3. The banner for the food fair is 5 yards long. The width of the
banner is $\frac{1}{3}$ of the length. How many feet wide is the banner?

1 yard is a measure of length.

1 yard is the same length as 3 feet.

Solution: _____

4. It takes Jen $\frac{1}{2}$ hour to walk to the market. Jen has walked $\frac{2}{3}$ of the
distance. About how many minutes has Jen been walking?

There are 30 minutes in $\frac{1}{2}$ hour.

Jen walks about 3 miles per hour.

Solution: _____

Mixed Numbers

Complete the chart.

Model	Mixed Number	Mixed Number (words)
1.		one and two thirds
2.		
3.	$2\frac{3}{4}$	
4.		three and one fifth
5.	$4\frac{1}{2}$	
6.		one and one third

7. Jan made 4 pies. She ate $\frac{1}{3}$ of a pie. How many pies does she have left?

_____

Use with Grade 3, Chapter 26, Lesson 1, pages 576–577.

Explore Adding Fractions

Why did it get hot in the stadium after the game?

To find the answer, add. Then write the letter from each exercise above the matching sum in the code.

1. $\frac{2}{10} + \frac{3}{10} = \frac{5}{10}$ **H**

2. $\frac{1}{5} + \frac{2}{5} =$ _____   **L**

3. $\frac{2}{6} + \frac{2}{6} =$ _____   **A**

4. $\frac{1}{8} + \frac{5}{8} =$ _____   **T**

5. $\frac{6}{12} + \frac{1}{12} =$ _____   **F**

6. $\frac{2}{8} + \frac{3}{8} =$ _____   **S**

7. $\frac{2}{9} + \frac{1}{9} =$ _____   **N**

8. $\frac{4}{10} + \frac{4}{10} =$ _____   **E**

Because __<u>   H   </u>__ ___ ___   ___ ___ ___ ___   ___ ___ ___ ___

$\frac{6}{8}$ $\frac{5}{10}$ $\frac{8}{10}$ $\frac{7}{12}$ $\frac{4}{6}$ $\frac{3}{9}$ $\frac{5}{8}$ $\frac{3}{5}$ $\frac{8}{10}$ $\frac{7}{12}$ $\frac{6}{8}$

Explore Subtracting Fractions

Here is a pattern with its rule:

$\frac{1}{2}$, 1, 1$\frac{1}{2}$, 2, 2$\frac{1}{2}$

Rule: Add $\frac{1}{2}$.

Complete each pattern. Write the rule beneath.

1. $\frac{1}{4}$, $\frac{3}{4}$, 1$\frac{1}{4}$, _____

Rule: _____

2. $\frac{7}{8}$, $\frac{5}{8}$, $\frac{3}{8}$, _____

Rule: _____

3. 1, $\frac{5}{6}$, $\frac{4}{6}$, _____, $\frac{2}{6}$

Rule: _____

4. $\frac{1}{10}$, $\frac{4}{10}$, $\frac{7}{10}$, 1, _____

Rule: _____

5. $\frac{9}{10}$, $\frac{7}{10}$, $\frac{5}{10}$, _____, $\frac{1}{10}$

Rule: _____

6. $\frac{11}{12}$, $\frac{8}{12}$, $\frac{5}{12}$, _____

Rule: _____

7. Create a new pattern with this rule: Subtract $\frac{1}{4}$.

_____

Use with Grade 3, Chapter 26, Lesson 3, pages 580–581.

Add and Subtract Fractions

Add or subtract. Write each answer in simplest form.

Color the answers equivalent to $\frac{1}{4}$, green.

Color the answers equivalent to $\frac{1}{2}$, blue.

Color the answers equivalent to $\frac{2}{3}$, red.

Color the answers equivalent to $\frac{3}{4}$, yellow.

What is flying in the sky?

_____

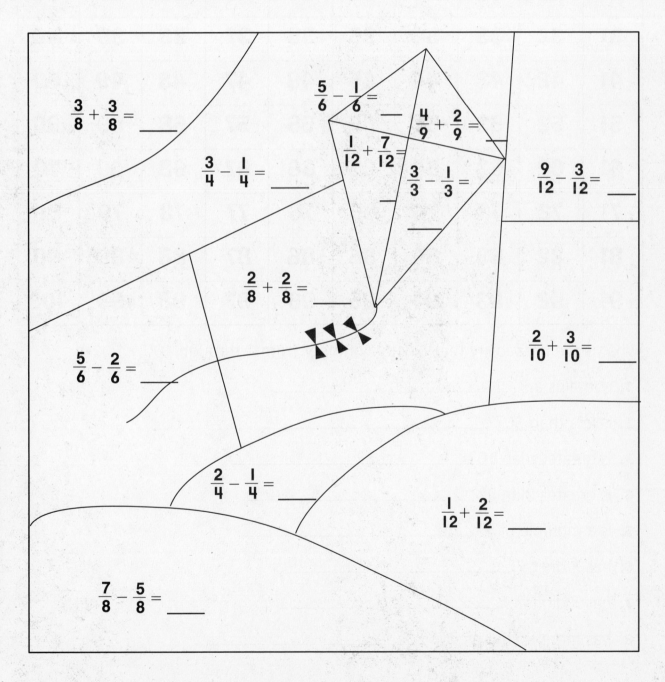

Probability

Use the hundred chart and the words *likely*, *unlikely*,
certain, or *impossible* to describe the probability.

1	2	3	4	5	6	7	8	9	10
11	12	13	14	15	16	17	18	19	20
21	22	23	24	25	26	27	28	29	30
31	32	33	34	35	36	37	38	39	40
41	42	43	44	45	46	47	48	49	50
51	52	53	54	55	56	57	58	59	60
61	62	63	64	65	66	67	68	69	70
71	72	73	74	75	76	77	78	79	80
81	82	83	84	85	86	87	88	89	90
91	92	93	94	95	96	97	98	99	100

What is the probability that a whole number from 1 through 100:

1. contains a 3? _____

2. is less than 30? _____

3. is greater than 10? _____

4. is greater than 0? _____

5. is a multiple of 5? _____

6. has 2 digits? _____

7. has 3 digits? _____

8. has more than 4 digits? _____

Use with Grade 3, Chapter 26, Lesson 5, pages 586–587.

Explore Finding Outcomes

1. Design an unfair spinner. Draw it here.

2. Design a fair spinner. Draw it here.

3. Invent a game that uses the fair spinner. Write the rules for your game here. Then make the spinner and play your game.

_____

_____

_____

_____

_____

4. Explain what makes a spinner fair or unfair.

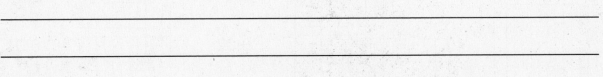

_____

_____

Explore Fractions and Decimals

Play a game with a partner. Cut out the cards below. Mix them up.
Give each player 4 cards. Put the rest of the cards facedown.

Try to match two cards that show the same amount. If you need a
card to make a match, ask for it during your turn. If your partner
doesn't have it, take the top card off the pile. The first person to lay
down all his or her cards in matches wins the game.

	0.4		0.2		0.55
	0.9		0.48		0.08
	0.8		0.33		0.25
	0.5		0.81		0.04
	0.1		0.02		0.01

Use with Grade 3, Chapter 27, Lesson 1, pages 606–607.

Fractions and Decimals

Match the decimals below to the word names, models, or fractions in the box. Write the letters on the line to find the name of each inventor.

1. I invented the first home video game in 1972.

R ___ ___ ___ ___   ___ ___ ___ ___
0.75 0.6 0.94 0.3 0.7   0.05 0.6 0.1 0.75

2. In 1960 I built the first working laser.

___ ___ ___ ___ ___ ___ ___ ___
0.5 0.7 0.1 0.64 0.09 0.64 0.75 0.1

___ ___ ___ ___ ___ ___
0.38 0.6 0.06 0.38 0.6 0.2

3. I invented videotape in 1950.

___ ___ ___ ___ ___ ___ ___ ___ ___   ___
0.6 0.94 0.1 0.01 0.6 0.2 0.09 0.1 0.75   0.38

___ ___ ___ ___ ___ ___ ___ ___ ___ ___ ___
0.3 0.64 0.06 0.2 0.5 0.06 0.6 0.5 0.64 0.04 0.04

4. How do you remember what the decimals 0.1 and 0.01 represent?

_____

_____

_____

_____

C = 9 tenths

D = 9 hundredths

E = 1 tenth

F = 4 hundredths

H = 7 tenths

I = 6 hundredths

K = 25 hundredths

L = 94 hundredths

M = 38 hundredths

N = 2 tenths

O = 64 hundredths

P = $\frac{3}{10}$

R = $\frac{75}{100}$

S = $\frac{35}{100}$

T = $\frac{5}{10}$

X = $\frac{1}{100}$

Decimals Greater Than One

Race for decimals with a partner. You will need a game board, scissors, a red and a blue crayon.

Cut out the number cards below. Mix them up and place them facedown. Player 1 takes a blue crayon and Player 2 takes a red. Player 1 turns over one card. He or she marks off on the game board the number shown on the card. Player 2 turns over a card and marks the game board where Player 1 left off. The winner fills in the last space on the game board.

Game Board

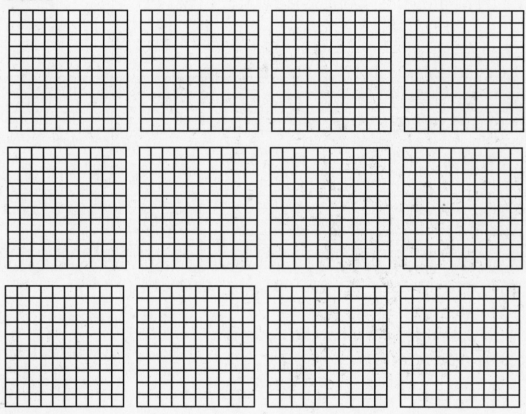

Number Cards

0.1	0.4	0.7	1.0	1.2	1.35	1.5	1.65	1.8	1.95
0.2	0.5	0.8	1.1	1.25	1.4	1.55	1.7	1.85	2.0
0.3	0.6	0.9	1.15	1.3	1.45	1.6	1.75	1.9	2.1

Use with Grade 3, Chapter 27, Lesson 3, pages 612–613.

Compare and Order Decimals

Follow the paths. Always follow the path of the greatest number when there is a choice of paths to follow.

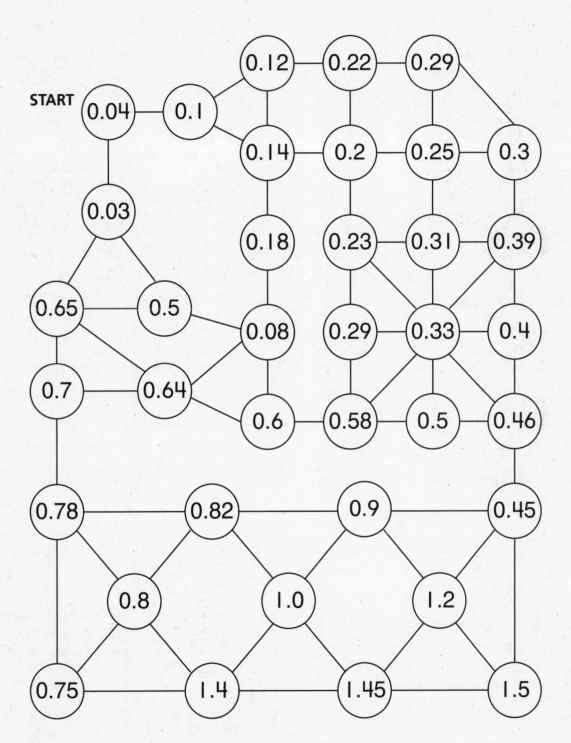

On which number did the path end? _____

Problem Solving: Skill
Choose an Operation

Solve. Tell how you chose the operation.

1. Lisa buys a book about inventors. The book costs $12.95. Lisa pays with a twenty-dollar bill. How much change should she receive?

_____

_____

_____

2. An inventor spends $345.15 on parts and $20.98 on paint. What is the total amount that the inventor spends?

_____

_____

_____

3. Vanessa spends $9.20 on supplies for her invention. Wendy spends $8.15. Who spends more? How much more?

_____

_____

_____

4. Harvey sells an invention for $27.95. Carmen sells an invention for $42.50. How much more money does Carmen receive than Harvey?

_____

_____

_____

5. Nina sells one invention for $59.75 and another invention for $89.95. How much money does Nina receive?

_____

_____

_____

6. Kareem spends $68.32 on supplies for his invention. He pays with a hundred-dollar bill. What is his change?

_____

_____

_____

7. Use the information in the table to write and solve your own problem.

_____

_____

_____

Year	Invention
1785	parachute
1885	bicycle
1939	helicopter

Source: *World Almanac for Kids*, 2000, World Almanac Books

Use with Grade 3, Chapter 27, Lesson 5, pages 618–619.

Name _____

Explore Adding Decimals

Question: What popular, nonpolluting vehicle was invented in 1885?

_____

Find the answer to the question above. You will need a pair of scissors and a roll of tape.

Step 1 Write the decimal represented by each model below.

Step 2 Cut out the puzzle pieces along the edge of the paper. Solve the problem on each piece. Arrange puzzles pieces in order by matching each answer with a model.

0.6 + 0.4 =

_____

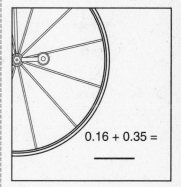

0.16 + 0.35 =

_____

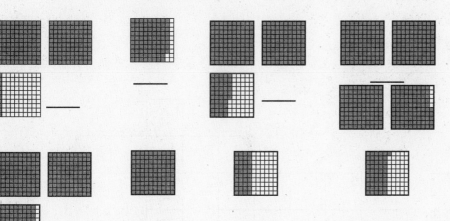

How is adding decimals like adding whole numbers?

_____

0.1 + 0.3 =

_____

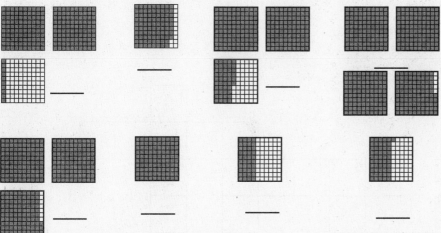

1.2 + 0.9 =

_____

1.06 + 1.40 =

_____

2.90 + 0.03 =

_____

0.38 + 0.50 =

_____

2.35 + 1.60 = _____

Use with Grade 3, Chapter 28, Lesson 1, pages 624–625.

Add Decimals

Add. Write the sums in the crossmath puzzle.
Write each decimal point in its own box.

Across

2. 6.34
    + 5.18

4. 8.75
    + 7.37

7. 4.76
    + 2.03

8. 4.19
    + 3.14

Down

1. 1.83
    + 1.18

2. 9.6
    + 2.7

3. 7.35
    + 1.28

5. 5.24
    + 1.68

6. 6.82
    + 1.93

9. 2.2
    + 1.3

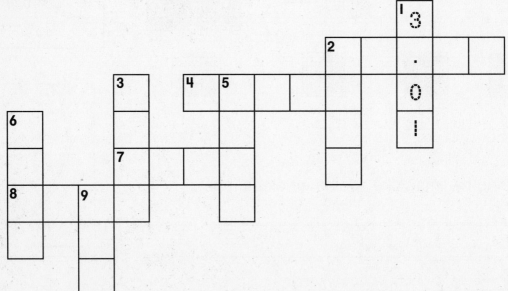

Why is it important to write the decimal point in the sum when you add
decimals? _____

_____

_____

Use with Grade 3, Chapter 28, Lesson 2, pages 626–628.

Name _____

Explore Subtracting Decimals

Subtract. Write each answer on the line.
You may use models if you wish.

D 2.4 − 1.8 = _____   **W** 3.25 − 1.60 = _____

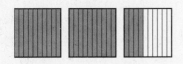   

T 0.75 − 0.28 = _____   **O** 4.63 − 2.45 = _____

G 1.4 − 0.6 = _____   **I** 3.50 − 1.25 = _____

N 0.52 − 0.45 = _____   **O** 2.70 − 2.62 = _____

O 5.1 − 0.4 = _____   **E** 3.98 − 3.28 = _____

Match each difference to a number in the answer to the riddle below.
Write the letter of each difference on a line to solve the riddle.

An inventor invented a wooden car. It had a wooden engine.
It had wooden wheels. What happened to the car?

__ __   __ __ __ __ __ __   __ __
2.25 0.47   1.65 4.7 2.18 0.6 0.7 0.07   0.8 0.08

Why do you need to regroup when you subtract 2.4 − 1.8?

_____

_____

_____

Subtract Decimals

Choose numbers from the box. Write your own subtraction
sentences so that you end up with the difference shown.

There is more than one way to make each subtraction
sentence. Use decimal models or a calculator if you wish.
You may use each number more than once.

0.08	0.1	0.16	0.7	0.8	1.0	1.08
	1.4	1.7	2.3	2.36	3.26	4.5

1. ___0.8___ – ___0.1___ = 0.7   2. _____ – _____ = 0.08

3. _____ – _____ = 3.1   4. _____ – _____ = 2.2

5. _____ – _____ = 0.3   6. _____ – _____ = 0.9

How did you decide which numbers to subtract to get a difference of 2.2?

_____

_____

_____

Use with Grade 3, Chapter 28, Lesson 5, pages 634–635.